Contents

Foreword

Welcome to Kew Gardens. There is so much here to enjoy and explore including our collection of nearly 20,000 different plant species, making it probably the most biodiverse spot on earth.

My aspiration for Kew Gardens is that every visitor will enter a magical world of botanical, architectural and artistic treasures. I hope you will be delighted and inspired by the beauty of our plants and landscapes and that you will learn about the critical role plants and fungi play in supporting all life on earth.

Kew is also a place of joy, tranquillity and relaxation. Science is telling us ever more about the many benefits to physical and mental health of spending time amidst nature.

Plants matter: all life depends on plants. They make food using energy from sunlight, carbon dioxide from the air and water. Healthy communities of plants and fungi provide humanity with the air we breathe, our food, medicines and fuel. They also help to regulate the climate and play a vital role in withdrawing carbon dioxide from the atmosphere, thus helping to mitigate climate change.

Behind the scenes Kew has an important global mission. We research and conserve the diversity of plant and fungal life on earth for the benefit of people and the future of all life on earth. Our aspiration is to end the extinction crisis and help to create a world where nature is protected, valued by all and managed sustainably.

Kew is a charity and supported by the Department of Environment, Food and Rural Affairs. We employ more than 300 scientists who work with partners in more than 100 countries around the world, researching and conserving plants and seeking to understand their many useful attributes. This work will help humanity solve some of our most pressing challenges – for example,

adapting agriculture to climate change by identifying heat and aridity tolerant crops.

Kew is also a place rich in history. Originally the private gardens of King George II and his son, Prince Frederick, and his wife Princess Augusta, the botanical collections started in 1759. We have a number of royal buildings at Kew including the iconic Great Pagoda, the palace and their kitchens and Queen Charlotte's Cottage. Some of the trees from this period are still with us today in the area close to the Orangery including the beautiful old ginkgo. Ever since these royal beginnings, Kew has been at the forefront of collecting, researching and displaying plants from around the world.

One of the great joys of Kew Gardens is how every visit is different. The changing seasons coupled with the scale offered by our 326 acres means that there is always something new to see and to discover. Despite working at Kew for more than a decade, I am struck how often a different plant, or a particular vista, catches my eye for the first time.

I hope you enjoy your visit.

Richard Deverell
Director (CEO)
Royal Botanic Gardens, Kew

Like the great River Thames that curves around it, the Royal Botanic Gardens, Kew is always changing. Since it was founded in 1759, it has driven ideas about science and nature and responded to emerging needs of communities around the world. Its landscapes and architecture have mirrored this change and its extraordinary plant collections continue to swell with species vital to the biodiversity of forests, mountains, deserts and savannahs.

But not everything changes here. Beauty is constant through the seasons and the years. Visitors experience first-hand the spectacle and wonder of plants in a unique setting. This Guide aims to enrich that experience by exploring some of the stories behind Kew's plants and people, past and present.

Welcome to Kew

Walking through the gates of the world's leading botanic garden, you discover a place where plants take centre stage. You are struck by the scale and beauty of the trees, which ring with bird song. Paths and sweeping lawns lead you to stunning architecture from every stage in the Gardens' history, built to display rich collections of plants, art and artefacts.

Throughout Kew's historic landscape, enchanting gardens, temples and statues await your arrival. Vistas, glades and dells offer you space to contemplate and to play, while the lofty chambers of the great glasshouses transport you to tropical forests and arid plains.

The Gardens embody the purpose of Kew as the global resource for plant and fungal knowledge. They bring science to life – stimulating curiosity, provoking debate, celebrating beauty and inspiring a love of the natural world.

THE BIRTH OF THE GARDENS

The Gardens were created as 18th-century Britain began to make sense of the world through reason and science. European explorers traversed the globe and brought back exciting plants. The two adjacent royal estates that were to become the Royal Botanic Gardens showed off these introductions as symbols of wealth and discernment (see Changing Landscape p118).

Sir Joseph Banks joined Captain Cook's famous expedition to the South Seas from 1768 to 1771, and on his return became Kew's first unofficial director, with the aim of creating the finest botanic garden of the age. Kew has been shaped by botanists and horticulturists over the centuries, who have continued to bring the rich variety of the plant kingdom to these Gardens, and expanded its extraordinary collections of objects and botanical art (see Art and Artefacts p82).

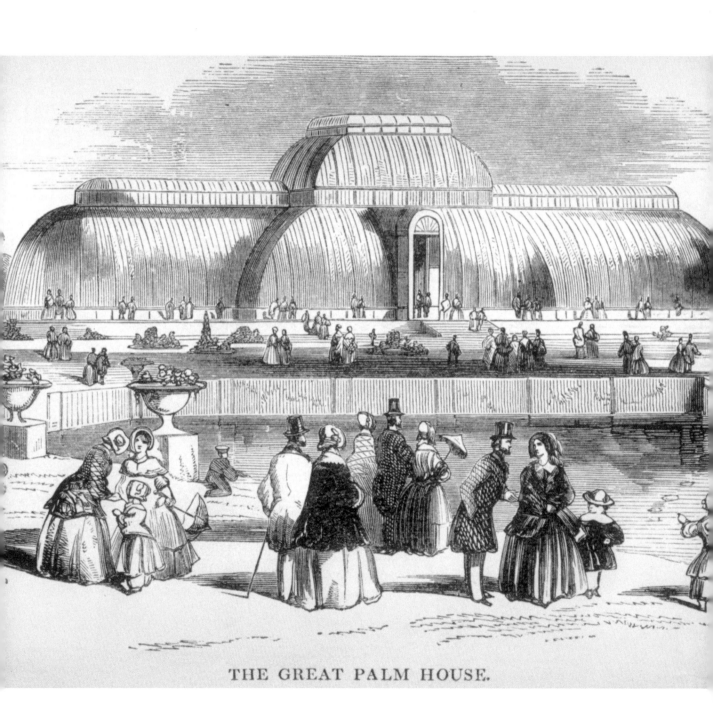

THE GREAT PALM HOUSE.

QUEST FOR PLANTS

In the 1800s, British colonies were being founded across the world. The hunger for novelty transformed into a quest for plants of economic value (see People Need Plants p24). In 1840, Kew was given by the Crown to the nation. The first official director, Sir William Hooker, developed the Gardens to welcome the public, who came in their thousands to marvel at the plants, displayed outside and in purpose-built glass palaces.

The Palm House and Temperate House are fantastic examples of cutting-edge Victorian technology and Imperial spectacle, and have set the standard for Kew's architecture to this day (see Architecture p92). Sir William Hooker was succeeded as director by his son, Joseph Dalton Hooker. One of the greatest British botanists of the 19th century, he made Kew the hub of a global plant exchange network. For example, seeds of the rubber tree, *Hevea brasiliensis*, were received at Kew in 1876. From these, seedlings were grown, which were then sent to Sri Lanka and Malaysia, becoming the basis for their massive rubber industries.

Joseph Dalton Hooker, director of Kew 1865–85, renowned Victorian scientist and great friend to Charles Darwin.

HORTICULTURAL EXCELLENCE

Kew has offered formal training to gardeners since 1859 and its alumni continue to enrich British and international horticulture. Many have gone on to be great horticulturists themselves, like graduate Ernest Henry Wilson, who introduced plants from China in the early 20th century, including some of the wonderful magnolias that grace the Gardens today (see Magnolias p58). Kew's horticulturists rise to the challenge of creating optimal growing conditions for 16,900 species of plants at our 132 hectare (326 acre) site (see Plants p10).

A Wardian case, developed in 19th century, used to keep plants alive and transport them around the world, spurring a revolution in the movement of plants.

SCIENCE AND CONSERVATION

While the power and potential of the world's plants are still being revealed, the threat of over-exploitation and environmental change is increasingly apparent. The Gardens are full of plants of economic and medicinal importance including oak, yew, aloe, pepper and bamboo, as well as those of celebrated beauty such as magnolia, tulip, orchid, waterlily and rhododendron. But among them is a growing number of species endangered in the wild (see Back from the Brink p14, and Living Fossils p68). In response, Kew's purpose has evolved; we now play a pivotal role in an international network of partnerships, as a world leader in plant and fungal science, conservation and plant-based solutions to global environmental challenges.

INSPIRATIONAL GARDENS

Since 2003, the Gardens have been a UNESCO World Heritage Site, recognising the unique combination of plants, architecture, art and landscape and over 260 years of science and ideas. And, Kew is still evolving, through the work of our scientists, horticulturists, educators and many volunteers. Kew reaches out to audiences of every age and origin to delight and inform them, and to inspire them to join us in caring for the natural world and forging a sustainable future for plants and people.

Autumn colours reflected in
the Palm House pond.

HOW KEW IS FUNDED

The Royal Botanic Gardens, Kew is a non-departmental public body under the National Heritage Act. About one third of Kew's annual running costs are met by the UK government through our sponsoring body, Defra (Department for Environment, Food and Rural Affairs). We generate the rest of our income from entrance fees, membership, commercial activity and philanthropic donations. The generous contributions made by our members and supporters are increasingly important to maintain the Gardens, and to support our world-leading science and conservation projects worldwide.

The Great Broad Walk Borders
planted for summer colours.

Plants

Kew's plant collections have long been renowned for being amongst the most valuable and varied. Home to an impressive 16,900 species of plants from all over the world this rich treasure trove of biodiversity was officially celebrated by the famous Guinness World Records book in 2022 as one of the greatest, with a new record: 'the largest collection of living plants at a single-site botanic garden'.

With horticultural skills and knowledge developed over 250 years, and the latest technologies, the glasshouses and gardens of Kew provide carefully managed growing spaces for plants, creating a living reference library for our scientists and a dynamic landscape in which to display the beauty, wonder and power of this diverse natural kingdom on which all life depends.

Passiflora phoenicea, one of the many species used for passion fruits, growing in the Princess of Wales Conservatory.

Plants Under Glass

Kew's glasshouses transport you to other worlds – to the steamy rainforest, the cool arid desert and the bright breezy mountainside. Throughout its history Kew has pioneered techniques to give its plants the growing conditions they need to flourish far from their native lands.

Growing plants under glass presents many challenges. Each species has particular requirements for humidity and temperature, food and light levels. They have come from complex ecosystems – some depend on fungi in the soil to access nutrients, others on insects for pollination or birds to disperse their seeds. They respond to wet and dry seasons, hot and cold.

Flowering can be triggered by the length of the day, seed germination by cold or by fire. To grow them successfully, Kew horticulturists must be observant, persistent, skilled and passionate about plants.

The elegant Palm House **K7** is devoted to the rainforest and includes plants that have changed the world. The Para rubber tree, *Hevea brasiliensis*, was the basis of a global industry, just as the African oil palm, *Elaeis guineensis*, is today. Kew research is helping to develop sustainable harvests of these and other products like cocoa and coffee, bananas and vanilla. Important crops like rice, lemon grass and sugar cane are often grown around the central pool in the intimate Waterlily House **K6** , home to the stunning Santa Cruz waterlily, *Victoria cruziana*, and the sacred lotus, *Nelumbo nucifera*.

The distinctive arched Davies Alpine House provides excellent conditions for bulbs and plants from mountain ranges and Arctic shorelines, keeping them dry in winter and cool in summer, while the Bonsai House **N8** , gives shade and space to contemplate its fascinating miniaturised trees. The Temperate House **E7** underwent a major five-year restoration, completed in 2018, and houses important collections from Americas, South Africa, Asia, Australasia and island temperate habitats. These many endangered species being conserved by Kew (see p107).

The Princess of Wales Conservatory displays an astonishing array of plants in ten climatic zones. Here you can find cacti, bromeliads and orchids as well as the titan arum, *Amorphophallus titanum*. Two zones are devoted to carnivorous plants including pitcher plants, *Nepenthes*, and Venus flytraps, *Dionaea muscipula*, which can close on an insect in under half a second.

Victoria boliviana, the largest waterlily in the world, native to Bolivia and a species new to science, first described in 2022 by a team of experts in science, horticulture and botanical art.

One of Kew's iconic plants grows in the wet tropical zone. The celebrated giant waterlily, *Victoria amazonica* is grown from seed each year. The leaves first appear as spiny heads but expand rapidly to reach up to 2.8 metres across. The spectacular white flowers open at sunset, filling the air with strong perfume. At Kew, the flowers are hand-pollinated, but in the wild scarab beetles are drawn to feed on the flowers, which then close at night. The next day the flowers turn pink, release their pollen and re-open. The beetles make their escape, brushing pollen onto their sticky bodies as they leave. That night they are drawn to the fragrance of another white flower and pollinate it as they land.

Back From the Brink

The stories of two plants on the edge of extinction
bring stark clarity to Kew's conservation work.

THE PYGMY WATERLILY OF RWANDA

Nymphaea thermarum, the
world's smallest waterlily.

The world's smallest waterlily, *Nymphaea thermarum*, was identified in 1987
growing in the mud around freshwater hot springs and is known from just one
location in Mashyuza, in south-west Rwanda. However, it disappeared from there
due to over-exploitation of the hot spring that fed this fragile habitat. Water
was prevented from reaching the few square metres where this species grew.
Nymphaea thermarum became extinct in the wild.

Thankfully, Professor Eberhard Fischer, the German botanist who had first
described it, realised that the species was under threat. He had transported
a few plants to Bonn Botanic Gardens in Germany, where the horticulturists
successfully grew these valuable specimens for more than a decade.

However, the pygmy waterlily proved extremely difficult to propagate and
it was feared that when these plants reached the end of their natural lives,
the species would be lost forever. As a result of a conservation plant exchange

with Bonn, a handful of plants came to Kew. After many trials, Kew horticulturists managed to replicate the damp warm conditions of the species' natural habitat and finally coaxed this tiny rarity into flower in 2009.

Once the riddle of its propagation had been solved, *Nymphaea thermarum* was easily cultivated and now many plants are growing at Kew, some of which can be seen in the Princess of Wales Conservatory.

Kew works with partners around the world to build consensus on conservation policy and makes an impact on the ground by offering training in conservation techniques and habitat restoration. Though it is now piped away before it reaches the surface, the hot spring water still flows beneath the native habitat of *Nymphaea thermarum*, and there may be an opportunity in future to restore the site and re-introduce this beautiful species to Rwanda.

CAFÉ MARRON, LOST AND FOUND

Café marron, *Ramosmania rodriguesii*, is a wild coffee relative endemic to Rodrigues, a remote island in the Indian Ocean. The evergreen shrub with glossy leaves and star-shaped white flowers was thought to be extinct. It was last seen in the 1940s, known only from dried specimens and a drawing made in 1877. But in 1984, when shown the drawing by his teacher, a 12-year-old boy insisted that it matched a plant growing near his house. Samples were sent to Kew and we confirmed that it was the long-lost café marron.

Careful exploration of the island failed to find another specimen and it became clear that this one plant was the last representative of the species on Earth. Measures were taken to protect the plant on the island but any hope for the future of the species depended on creating new plants.

Kew horticulturists managed to root cuttings sent from Rodrigues but it was another 20 years before they worked out how to produce fruit. In 2003 the first berry ripened, producing the seeds of new individuals and in 2009 Kew was at last able to send healthy plants back to their island

home, where work is ongoing to re-establish a stable population in the wild. At Kew they are grown in the Palm House and the Princess of Wales Conservatory.

The story of café marron shows the vulnerability of island floras. Due to their isolation and restricted size, islands are often rich in endemics – plants and animals that occur nowhere else in the world. The natural balance of island ecosystems can be easily upset by the introduction of non-native species. Without any native grazing animals on Rodrigues, its plants had no features to deter the goats that were brought to the island by settlers and grazed café marron to the brink of extinction.

Rodrigues is part of the Republic of Mauritius, where 275 species of plants are critically endangered. Specialist horticultural skills developed at Kew are needed now more than ever to replicate the positive stories of the pygmy waterlily and café marron elsewhere in the world.

The flowers of the small evergreen café marron shrub.

Once the widest single-span glasshouse in the world, the Waterlily House was built to house tropical waterlilies and climbing plants.

Perfect Adaptations

Plants have developed in their natural environments over millions of years and some of their extraordinary adaptations can be seen among plants in the Princess of Wales Conservatory.

Charles Darwin's theory of evolution by natural selection was hugely influenced by the variety of plant adaptations he saw, and he shared his radical ideas with his friend Sir Joseph Hooker, Kew's second official director, as letters in Kew's archive show (see Art and Artefacts p82).

Climbing the column by the stairs in the arid zone of the Princess of Wales Conservatory **N7** is a cactus that is adapted to attract bats. The pitaya or dragon fruit cactus, *Selenicereus costaricensis* comes from tropical forests of Central America and as far south as Peru. Its buds open at dusk into wide cupped pale flowers that emit a sweet smell and are thought to reflect the sound signals of a hungry echo-locating bat. The bat forces its head into the flower, trying to reach the nectar with its long tongue. In the process its furry body collects large amounts of pollen, which it transfers to the next flower it visits.

Left: *Nepenthes truncata*

GIANT OF THE RAINFOREST

Another night-flowering plant is the titan arum, *Amorphophallus titanum*, from western Sumatra. But its massive flower structure produces neither nectar nor sweet fragrance; on the contrary, it emits the nauseating stench of rotting flesh, an adaptation that attracts insects from great distances in the tropical forest. Every year the single immense mottled leaf of the titan arum grows to the size of a small tree before dying down again. This annual growth builds up food reserves in its underground tuber until it can produce its huge flower, which rises up to three metres above the ground. The first time it flowered in cultivation was at Kew in 1889, exciting great public interest. In 1926, when it flowered again, the crowds were so large that the police were called to control them. Kew horticulturists have mastered its cultivation and there is almost always a plant in leaf in the wet tropical zone, where the high temperature, shade and humidity recreate the conditions of its natural habitat. The plant can take 20 years to flower from seed and the tubers have been known to weigh up to 90kg, which makes repotting a challenge.

Plants have adapted in many ways to find nutrition in inhospitable habitats. Carnivorous plants trap prey such as midges and flies to get nutrients, and their different mechanisms can be seen in two dedicated zones and in the wet tropical zone. Though temperate *Sarracenia* and tropical *Nepenthes* are unrelated, they have evolved to catch their food in the same way and both have pitchers with smooth waxy interior walls. Insects are attracted to the colours and sweet secretions inside the pitchers but once inside they slip, drown and are then digested.

Right: The huge flower of the titan arum which rises up to three metres above ground and emits a nauseating smell of rotting flesh.

Encephalartos woodii (foreground) in the Temperate House.

Cycads

Kew has a fine collection of cycads – plants that were widespread on Earth over 250 million years ago, before the dinosaurs and well before the appearance of the flowering plants that now dominate the world's vegetation.

Some of the rarest plants in the Gardens are cycads. With their stout trunks and dense crowns of large whorled pinnate leaves, cycads resemble palms or tree ferns, but they have many unique characteristics. They are long-lived, slow-growing plants that produce male and female cones in the centre of the leaf whorl on separate plants.

One of the cycads at Kew is also a contender for the title of oldest pot-plant in the world. Housed in a large wooden box at the southern end of the Palm House **K7**, the Eastern Cape giant cycad, *Encephalartos altensteinii*, arrived at Kew in 1775, sent from South Africa by Kew's first plant hunter, Scottish botanist Francis Masson. In 1819, 44 years after it came to Kew, the Eastern Cape giant cycad produced its first cone. It has not produced another cone since. Now weighing more than a tonne, this venerable plant measures over four metres and leans languidly sideways in a sign of its old age. Though classed as vulnerable in the wild due to the pressures of land clearance, the species is fairly common in South Africa, growing in open shrubland on rocky slopes and in evergreen forest.

You can see many cycads in the Palm House including *Encephalartos ferox*, which produces bright orange-red cones. Pollinating cycads is a tricky business but well worth the effort as they are so endangered in the wild. Kew's 90-year-old male *Encephalartos ferox* hit the headlines when its pollen was extracted using a turkey baster to fertilise a female cycad in Bristol.

In the Temperate House you can see one of the world's rarest cycads, *Encephalartos woodii*, often called the world's loneliest plant. Only one specimen (a male) was ever found in the wild in South Africa. Kew's specimen is an offset of that original plant and arrived here in 1899.

Cycads are in high demand for the architectural drama they bring to gardens, but to protect wild populations their trade is highly regulated. Kew works internationally, advising governments and assisting the UK and other countries to set priorities for conservation and sustainable use of plant resources, in order to ensure a positive future for plants like the great cycads.

Orchids

Kew has one of the best collections of tropical orchids in the world. Many are epiphytic – growing in the wild on the trunks and branches of trees. Kew horticulturists aim to mimic their environment in cultivation, providing the right temperatures, humidity and growing places to keep the orchids happy.

The Princess of Wales Conservatory **N7** has two distinct zones devoted to the display of these beautiful plants. A hot, steamy zone features the tropical lowland species with their spectacular showy flowers and adaptations for life in the forest canopy. A second, cooler orchid zone suits the no less attractive species from tropical mountain regions. Many more specimens in Kew's collection are grown in controlled environments in the Tropical Nursery and put on display when they come into flower. You'll find perfumed blooms of *Cattleya and Dendrochilum,* strange-looking *Paphiopedilum* slipper orchids, and butterfly-like *Dendrobium*, as well as long branching sprays of small bright flowers of *Oncidium*, tall spikes of the waxy blooms of *Cymbidium*, and the unusual deep-blue and burnt-orange colours of flamboyant *Vanda*. There are also many miniature orchids, with flowers only a few millimetres across.

Technological advances mean orchids can now be produced commercially on a huge scale, with plant breeders developing ever more amazing hybrids and cultivars. Moth orchid, *Phalaenopsis,* hybrids are now the most popular house plants in the world. However, many species are endangered in the wild by habitat destruction and over-collecting. Kew, a world leader in the study of orchids, is working to help ensure the conservation and sustainable use of these plants in their natural habitats.

Orchids inspire obsession among scientists and gardeners who have discovered, studied and recorded them for centuries. At a time of orchid mania in Victorian Britain, Charles Darwin published a book on orchids, using his ideas about evolution to explain why they produced such elaborate flowers. The orchid family, Orchidaceae, is one of the largest of all plant families, with around 28,000 species – one in twelve of all flowering plants – distributed across the world. 3,500 orchid species have been recorded in Ecuador while only 55 species – all terrestrial – grow naturally in Britain.

Kew celebrates this fascinating plant family at its annual orchid festival in late winter, when thousands of brilliant flowers are put on display.

Oncidium naevium in the Princess of Wales Conservatory.

Preparation for the annual orchid festival in the Princess of Wales Conservatory.

People Need Plants

In Kew's magnificent Palm House, the heat is high, the air is heavy with humidity and sunlight filters through multiple textures and shades of green leaf. These are perfect growing conditions for plants from tropical rainforests, many of which have huge economic and medicinal value.

Rattans are a large group of extremely spiny climbing palms that are a focus of research at Kew, and several species grow in the Palm House **K7** . The central part of the long stem of a rattan forms a solid cane, which has a wide range of uses including furniture making and handicrafts and on which a multibillion-dollar worldwide trade has been built. The longest plant stem ever recorded was a rattan, a species of *Calamus*, that measured nearly 200m in length.

In the Palm House you can see the African oil palm, *Elaeis guineensis*, which gives us the most productive oil crop in the world. Palm oil, extracted from the fruits and seeds, has an enormous range of different uses including fuels, cosmetics and food. However, much of the destruction of forests in some of the most biodiverse areas of the world has been blamed on the extensive cultivation of this species, often in vast monocultures.

Enset, *Ensete ventricosum*, the so-called false-banana, is being explored by Kew and its partners as a crop to help fight rising food insecurity across Africa. The stem and roots of the plant are the edible parts, not the fruit. They are used to make a bread-like food, kocho and bulla, a white powder for making pancakes and dumplings.

In the Palm House you can see a variety of rainforest plants of value to societies for the fruit, timber, spices and medicine they produce.

MEDICINAL POWERS

The pretty flowers of the rosy periwinkle, *Catharanthus roseus*, catch the shifting light beneath the Palm House trees and it has been a popular bedding plant for a century. But it is also known as a useful folk medicine and modern science found it to be rich in powerful alkaloids. These are now used in treatments for cancers, in particular childhood leukaemia and non-Hodgkin's lymphoma. The rosy periwinkle comes from the forests of Madagascar, which have been heavily damaged by human activity, but the plant thrives in disturbed areas so it has survived in its island home. Its story shows how important it is to preserve areas of rich biodiversity, for there may be other 'miracle drugs' in plants yet to be discovered.

FOOD CROPS WE DEPEND ON

Banana plants are one of the most economically important crops in the world, providing food, shelter and fibre. Thousands of years of domestication have produced the delicious fruit eaten by millions of people. You can see cultivars of the most important dessert bananas, *Musa acuminata* 'Dwarf Cavendish,' growing in the Palm House. It looks very different from wild relatives growing nearby, and other species which have green, silvery, striped or spotted fruits. Edible bananas contain no seed, so they can only be reproduced by separating suckers from the parent plants. This means they lack genetic diversity and are vulnerable to decimation by pests and diseases, as happened in the 1950s when the market-leading banana was virtually wiped out by a single fungus. To prevent the repeat of this kind of disaster, Kew has worked with the Global Crop Diversity Trust on a ten-year project to protect, collect and prepare seed of the wild relatives of 29 key food crops, including the banana. They can then be used for the development of new varieties that are more resilient to the effects of climate change and more resistant to disease.

In addition to visual wonders like the titan arum and giant waterlily the Princess of Wales Conservatory **N7** houses plants of great significance to people. Food lovers can see the handsome cacao, *Theobroma cacao*, with its large seed pods housing the main ingredient for chocolate, and the delicious pineapple *Ananas comosus* – the only edible member of the spiky, highly ornamental bromeliads.

In the extensive dry tropics zone, growing among a spectacular array of cacti, are several species of *Agave*, which have long been used for fabrics, fuel, medicine and alcoholic drinks. These striking, architectural plants come from Mexico and northern South America. There are over 130 species, which have spiny, fleshy leaves adapted for survival in hot arid deserts and mountains. The long, tough leaf fibres of sisal, *Agave sisalana*, are used to make hard-wearing carpets, while the succulent leaf base of *Agave tequilana* is pulped and fermented to make tequila, the alcoholic drink.

Cocoa pods (*Theobroma cacao*) growing in the Princess of Wales Conservatory.

DEFENDING COASTAL COMMUNITIES

In the wet tropics zone by the central pool, slender grey-brown stems rise above the mud. These are the stilt roots of the red mangrove, *Rhizophora mangle*, adapted to withstand total submersion in salt water and the action of tidal waves. One of numerous mangrove species that colonise tropical and subtropical coastlines, red mangrove grows closest to the sea, forming unique ecosystems that have an increasingly important role as rising sea levels threaten tropical coastlines. They act as barriers to coastal erosion, provide natural 'nursery grounds' for many species of fish and other marine life, and trap sediments, allowing development of healthy off-shore coral reefs. As well as protecting coastal communities and providing good fishing, red mangrove is used for house building and canoes, fuel, dye and medicine. Although mangroves can cope with the stress of salt water, this is not a requirement, which is why their intriguing complex structures can be seen at Kew.

Mangrove (*Bruguiera gymnorhiza*)
flowering in the Princess of
Wales Conservatory.

The dry tropics zone in the
Princess of Wales Conservatory.

Plants in Gardens

Kew's mild climate enables plants from all over the temperate world to grow happily outside.

Our horticulturists use an abundance of different aspects – sunny walls, cool dells, woodland margins and watersides – to create the optimal growing environments. Kew is one of the sunniest places in the British Isles and July and August are the warmest months, with an average maximum temperate of 24°C. While January and February are the coldest months, hard frosts are infrequent. This means Mediterranean plants, including many salvias, grow to perfection in our free-draining soil. Olive and cork oak groves shelter the aromatic shrubs in the Mediterranean Garden (see p37), while south-facing walls protect the more tender salvias and lavenders in the north end of the Gardens. Salvias are important both for horticulture and for medicine and the beautiful salvia border by the Rock Garden  gives Kew scientists a living reference collection to investigate their potential in the treatment of dementia and cancer.

Salvia in The Great Broad Walk Borders (also pictured, right).

GARDENS WITHIN GARDENS

Some plants are grown according to their physical similarities – extensive brick walls at the northern end of the Gardens provide support for many climbing plants (see p46) – while other collections are grown to reflect their wild habitats in mountains and woodlands (see p34–35) or to display the astounding variety within one family, such as the grasses (see p48). Throughout its history, the Gardens have been at the vanguard of horticultural fashion.

Several collections are planted to recall gardens of different periods and cultures, such as the Queen's Garden **O4** , the Duke's Garden **P8** and the Japanese Gateway **B7** , Chokushi-Mon (see p44).

Below: The laburnum pergola in the Queen's Garden.

Above: The Japanese Gateway,
Chokushi-Mon.

WATERING THE PLANTS

One of the biggest challenges is keeping our plants well watered. South-east England is one of the driest parts of the British Isles with less than 650mm (26 inches) of rainfall throughout the year. While many plants receive a third of our rainfall in the wild, plants in southern China receive over three times more. Our plants from China, grown in many parts of the Gardens, have played a key role in Kew's project to identify and authenticate the ingredients in traditional Chinese medicine, so the health of the living collections is critical. Huge effort is made to improve our soils so they can hold water and nutrients throughout the drier months. Grass cuttings, tree and flower prunings are shredded and turned in enormous piles to produce the mulch which is then redistributed on our shrub and flower beds.

Left: Watering in the
Palm House.

Habitat Gardens

Mountains make up a quarter of the global landmass and mountain flora is as diverse as it is beautiful.

The Rock Garden and Woodland Garden, together with the Davies Alpine House **08** recall the habitats of temperate mountains, from deciduous forest to rocky outcrop. Here we grow around 4,000 herbaceous plants and shrubs, adapted to survive extreme weather conditions, perilous cliffs and shady clefts. The Alpine Nursery, behind the scenes, is home to another 3,000 plants, grown in pots, including National Collections of tulips and juno irises. These are displayed in the Davies Alpine House when in full flower.

LIFE ON THE EDGE

Alpines are plants that have adapted to live in harsh environments at high altitudes where trees cannot survive. They often have dramatic flower displays, out of proportion to their diminutive size, to attract insect pollinators that are fewer in number in the high mountains. A common survival strategy is a mat of hairs on leaf surfaces, providing protection from both the wind and the sun, which is more intense in the cleaner, thinner air.

First built in 1882 using limestone rock to mimic a Pyrenean mountain valley, complete with cascades and meandering streams, the Rock Garden is now constructed entirely of West Sussex sandstone, which supports a wider range of plants. It is planted in six geographic regions and while it reaches its flowering peak in May there are blooms to admire all year, right through to the miraculous snowdrops from the Taurus mountains in Turkey, *Galanthus elwesii*, which emerge just before Christmas.

The Woodland Garden.

JEWELS OF THE FOREST FLOOR

The high point of the Woodland Garden is in March and April. The warming sun reaches the forest floor, which bursts into flower before the trees come into full leaf. Many herbaceous woodlanders have developed bulbs or other underground storage organs to survive the dry and shade of summer. One such is the curious yellow wake robin (*Trillium luteum*), from the forests of the Appalachian Mountains of North America, whose three twisting petals sit on a platform of three mottled leaves. *Trillium* belong to the family Melanthiaceae, which have huge amounts of DNA in their cells. They are attracting keen interest from Kew's geneticists who are trying to understand the role that DNA plays in the way plants evolve.

Erythronium, lighting up the Woodland Garden in spring.

MEDITERRANEAN DIVERSITY

Five regions around the world share the distinctive climate that we call 'Mediterranean', with dry summers, mild moist winters and only sporadic frosts. These rocky coastal areas in California, Chile, South Africa, south-west Australia and the Mediterranean Basin are floral treasure-chests, home to 20% of the world's plant life – almost 50,000 species – though they cover less than 2% of its landmass.

Kew's Mediterranean Garden **H6** echoes the wild hillsides of southern Spain and the Greek islands. Many plants from the region have small evergreen leaves with leathery cuticles, to protect them from the hot sun and drying winds. Some species contain aromatic chemicals to deter herbivores like goats and rabbits. On warm days rich scents from these pretty shrubs such as lavender, *Lavandula*, and rock rose, *Cistus*, rise to the Tuscan portico of King William's Temple, which sits on a low mound above the garden. From here the view is framed by groves of trees that have played key roles in Mediterranean civilisations including the cork oak, *Quercus suber*, whose rugged thick grey bark is harvested for cork bottle stoppers for the wine industry; the stone pine, *Pinus pinea*, source of pine nuts and the olive tree, *Olea europaea*, whose small black fruits are rich in oil.

Lavender, *Lavendula*, yields rich scent on warm summer days.

CONSERVING ISLAND FLORA

For over a decade, Kew has been working with collaborators across Europe to conserve native and threatened plant species. As with all Kew's scientific collections, the Mediterranean Garden aims to showcase only 'natural sourced' material – plants grown from seed collected in the wild, so that this beautiful place continues to be the most magnificent living reference library in the world.

The Mediterranean Garden.

Spring tulips and blossom along Cherry Walk by the Temperate House.

Designed Gardens

Kew Gardens has changed throughout its history, reflecting the development of its purpose and changing ideas about how plants should be displayed.

While the major influences are from the 18th and 19th centuries, garden styles of other periods and cultures can be found in secluded corners. The Queen's Garden **04** complements Kew Palace, drawing on 17th-century fashions in plants and garden architecture. Most of the plants seen here are those which were grown in Britain in the 1600s and earlier. The geometric parterre with its clipped box and fountain, the pleached hornbeam alley, mount and statues were all elements of the gardens of grand houses of the period. Scholars and physicians of the time published 'herbals' – compendia of plant lore and botanical cures. While many appear archaic and comical, they were often grounded in fact. It was said that common sage, *Salvia officinalis*, 'quickeneth the senses and memory' and, 400 years on, Kew scientists are investigating it as a treatment for Alzheimer's disease.

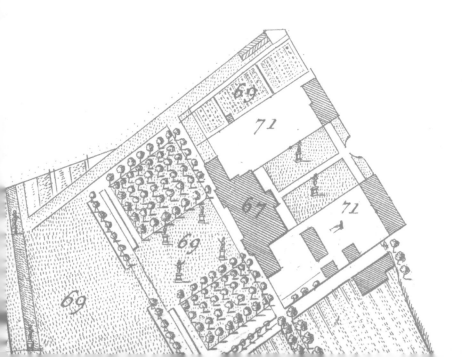

EVOKING THE EAST

An early garden style lies at the southern end of the estate, where late 16th-century Japan is evoked by small tea gardens of Peace, Activity and Harmony enclosing a symbolic miniature landscape, designed by Professor Fukuhara of Osaka University. On a raised mound at its centre is Chokushi- Mon, a replica of the Gateway of the Imperial Messenger in Kyoto **B7** . The finest example of a traditional Japanese building in Europe, it came to Kew after it was constructed for the 1910 Japan-British Exhibition in London. Look out for beautiful flowering cherries here in spring, anemones and rhododendrons in summer and stunning leaf colours in autumn.

THE GREAT BROAD WALK BORDERS

Planted over the course of 2015, the Great Broad Walk Borders N6 first flowered in 2016 to great acclaim. At over 320 metres long they are thought to be the longest double herbaceous borders in the country, possibly the world. You'll find beautiful spring bulbs here including alliums, daffodils and tulips, and plenty of autumn interest too, but these borders are at their peak in the summer months. They contain around 30,000 plants within eight large circular beds, each of which has a theme, showing off different types of plants or plant families.

ENGLISH FLOWER GARDEN

The walled Duke's Garden P8 has the more intimate atmosphere of a traditional 20th-century English flower garden. Its extensive borders are filled with bulbs and herbaceous perennials set off by large, manicured lawns while winter interest comes from fragrant shrubs. More contemporary is the Gravel Garden P8 , developed in the late 20th century in response to growing awareness about climate change. It displays attractive drought-tolerant plants that require minimal watering during summer, such as grasses, euphorbias and Mediterranean herbs.

The northern end of the Gardens has a wealth of mellow brick walls and other structures offering wonderful displays of climbing plants. Species from many different plant families have evolved to climb, using other trees and shrubs, rocks and cliffs, to reach sunlight without having to invest their energy into developing self-supporting structures.

The Duke's Garden.

Autumn in the Japanese gardens.

AGIUS EVOLUTION GARDEN

Located on the site of the old Plant Family Beds, the Agius Evolution Garden **N8** displays plants arranged according to the Plant Tree of Life. This classification is the result of research into plant DNA, revealing how plants are related to each other and how they have evolved over millions of years. A network of paths leads visitors through the planting, and low yew hedges separate the major plant groups, creating a beautiful and enlightening garden.

WISTERIAS AND ROSES

Wisterias and roses are among those climbers that develop woody frameworks that persist throughout their winter dormancy.

Wisterias are powerful vines from China and North America, which climb by twining their stems around any adjacent support. Kew's huge old *Wisteria sinensis* **07** , looks its best in April when its purple racemes cascade over an old iron pergola. You can sit inside this plant, admiring the fact it is nearly 200 years old. One of the first in Europe, it was propagated from a plant sent in 1816 from a Cantonese merchant.

Though the rose is seen as an iconic English plant, it has been cultivated in China for thousands of years, and in Europe since Roman times. The 120-metre-long Rose Pergola **N8** supports 26 cultivars of rose, which climb using thorns as grappling irons. Chosen for their profusion and length of flowering they peak in June creating a fragrant tunnel of colour.

Rose Pergola

Grasses

When grown en masse, the beauty and diversity of grasses is revealed. The areas at Kew devoted exclusively to this important plant family are striking and atmospheric.

The Bamboo Garden **H3** was created in 1892 in a secluded dell near the river, fuelling a craze among gardeners, who still delight in their elegant coloured stems, and leaves whispering of far-away places. In recent decades other grasses have become as popular in horticulture and the Grass Garden **O8** shows off their elemental grace.

Situated just north of the Davies Alpine House, it is designed to display a multitude of textures and colours, changing through the seasons.

Grasslands cover a fifth of the world's land and are absent only in parts of Greenland and Antarctica. They are intertwined with mankind's development, providing us with grains such as wheat, barley and maize, pasture for our livestock and shelter. Grasses are used for biofuel, building materials, paper and alcoholic drinks. The Grass Garden shows tropical and temperate cereal cultivars, while in the Palm House you can see tropical grasses such as sugar cane, *Saccharum officinarum,* which produces around 75% of the world's sugar supply.

HIGHLY EVOLVED

Grasses are wind-pollinated, so they have not evolved to produce showy flowers or strong scents to persuade insects or other creatures to do the work. They have hollow stems and grow from the plant base, rather than the tip, so that when they are grazed by herbivores they can regenerate – this is why cutting your lawn does not kill the grass. In spite of these adaptations, some grasses, including bamboos, are threatened in their wild habitats.

Kew scientists working on the remote South Atlantic island of St Helena, a UK Overseas Territory, identified a previously unnamed species of wavy hair grass.

The conservation of St Helena grass, *Eragrostis episcopulus*, is important, as the sites where it grows are rich in other rare plants. By retreating to remote cliffs and rocky outcrops, these pockets of diversity – found nowhere else in the world – have survived competition from non-native plant invaders and rabbits introduced by settlers. As a first step in the long-term conservation of the grass, seeds have been collected and stored in Kew's Millennium Seed Bank and the grass is now being grown in nurseries on St Helena and at Kew.

There are around 1,000 species of bamboo, growing in many different climates from cold mountain regions to warm, tropical areas, all over the world from China, Japan and the Himalaya to the Americas. They range from low-growing species to giant varieties that reach 35 metres tall and they are used to make scaffolding, house-frames, musical instruments and textiles.

The Grass Garden offers a chance to encounter the full beauty and variety of about 290 species of grass including *Bromus interruptus*, extinct in the UK until reintroduction in 2004.

Trees

Trees have been at heart of the Royal Botanic Gardens, Kew since its earliest days.

Several great specimens have witnessed its entire history, through agricultural and industrial revolution, world wars, the discovery of penicillin and DNA to the turn of the Millennium. In today's technology-driven, hyper-connected world tree species new to science are still being described and the value of the world's forests climbs ever higher, as we understand more clearly their contribution to the environment, climate stability, biodiversity and humanity.

Kew's hardy trees and shrubs combine in a living landscape that changes through the seasons. To explore them is to gain an insight into the beauty and diversity of forests around the globe and also, through the Natural Areas, to connect with a very local woodland.

The Arboretum

Kew's 12,000 magnificent trees are planted throughout the Gardens, representing over 2,000 species and varieties.

This great scientific collection provides grand vistas, glades of mixed woodland, `Interest weaves through the changing seasons, from dramatic structures laid bare by winter to the exquisite blooms of magnolias (see p58) and dogwoods in spring, and the mass displays of autumn colour by the maples, oaks and hickories.

Princess Augusta founded the Gardens in 1759 with five acres of trees planted according to the botanical understanding of the time. Within 15 years, the collection had swollen to nearly 800 species, including rarities like the monkey puzzle, *Araucaria araucana* **O6** , from Chile and the maidenhair tree, *Ginkgo biloba* (see p71). The botanic gardens later incorporated the adjoining royal estate, which had its own unusual trees such as the cedar of Lebanon, *Cedrus libani*, sweet chestnut, *Castanea sativa*, the tulip tree, *Liriodendron tulipifera,* and red oak, *Quercus rubra* (see p63). A wonderful specimen of the North American black walnut, *Juglans nigra*, known to have been planted in the 1730s, grows in the Woodland Garden **M8** .

Today's Arboretum – a 19th-century term derived from Latin and literally meaning 'a place with trees' – was developed by Sir William Hooker, who was director of the Gardens from 1840 when it was given to the nation by the Crown. In 1844 he challenged landscape designer William Andrews Nesfield to come up with a design for a scientific and educational tree collection that had a 'park-like character' (see Living History, p120).

The Great Pagoda and Syon Vista are features from this period and, with Cedar Vista, planted in 1871, form a monumental triangle of great avenues through the Arboretum, connecting the Palm House, the Pagoda and the River Thames at Syon House. Cedar Vista passes through the Woodland Glade **D5** , which are planted with collections of trees and shrubs from Kew expeditions, while Syon Vista is a stately progression of holm oaks, *Quercus ilex*. The original trees forming Nesfield's Pagoda Vista have been replaced by the North American pin oak, *Quercus palustris*, sweet gum, *Liquidambar styraciflua*, and Turkish hazel, *Corylus colurna*, which are suited to avenue planting and Kew's dry soil, and produce fiery displays in autumn.

The Treetop Walkway circles at 18m above the ground, giving the visitor not only breathtaking views across the Gardens and beyond, but also insights into the rich ecosystems of the tree canopy. From this lofty platform the seasonal rhythms of bird and insect life can be observed among the branches of sweet chestnut, *Castanea sativa*, beech, *Fagus sylvatica*, horse chestnut, *Aesculus hippocastanum* and several different oak species, *Quercus*.

Sir William Hooker's son, Sir Joseph Hooker, continued to develop the Arboretum, planting many trees and shrubs in groups according to genera for comparison and ease of research. Holly Walk **H6** and the Berberis Dell **H8** , laid out in the 1870s, show the variety of foliage, form and fruit of holly, *Ilex*, and barberry, *Berberis*, while the Redwood Grove **C4** , planted a decade earlier on the fringe of the Pinetum **C4** , is now home to Kew's tallest tree (see Conifers p64).

Horticulturists in the Arboretum are working to identify future trees for our changing climate. By collecting data from specimens in the Arboretum on how drought-tolerant trees may or may not be, they are developing an evidence-based approach to future tree selection.

MANAGING THE ARBORETUM

Today, Kew's Arboretum is managed by a permanent team of tree specialists, which works year-round to monitor and maintain the health and structural integrity of more than 12,000 trees. Highly skilled climbers make regular inspections, reaching the most inaccessible canopies using ropes and pulleys, while another team specialises in planting, ensuring that young trees get the best possible start in the Gardens.

Older trees, dating from the original royal estates or the inception of the Arboretum receive particular care. The soil around their roots is injected with a mixture of air and beneficial fungi to reduce compaction, allow them to absorb more nutrients, and increase their lifespan.

Fungi are vitally important to all plants, and have a well-known relationship with trees, forming associations with their roots called mycorrhizas. 90% of plants on Earth cannot survive without fungi in their roots.

Magnolias

Magnolias are adored for their velvet buds, extravagant blooms, handsome foliage and intense fragrance. With over 250 magnolias planted around the Gardens, springtime at Kew is synonymous with this wonderful genus.

There are around 300 species of these flowering trees and shrubs, native to East Asia and the Americas. Though there are tropical species, Kew's collection comes entirely from temperate regions. The first magnolia ever seen in Britain was the swamp bay magnolia, *Magnolia virginiana,* from North America, which was sent to Henry Compton, Bishop of London, in 1687.

An early addition to the Gardens at Kew was a cucumber tree, *Magnolia acuminata,* planted in the late 1700s, still growing on the Orangery Lawn. It is striking for its mini cucumber-like fruits which turn from green to bright pink in autumn before ripening to dark-red and splitting to release their red fruit.

Half a century after the North American species arrived came the first species from Asia. Sir Joseph Banks, Kew's unofficial first director, introduced the Chinese lily tree, *Magnolia denudata*, in 1780. By then it had been cultivated in China for at least a thousand years and valued as a symbol of candour and purity.

Most prolific of magnolia collectors was Ernest Henry Wilson, who had been a student gardener at Kew before embarking on a celebrated botanical career. In the early 20th century he introduced eight new species of magnolia that are still widely grown today, including one which bears his name, *Magnolia wilsonii*.

Magnolias are thought to have evolved early in the lineage of flowering plants. As far back as 65 million years ago they grew in vast forests together with sweet gum, *Liquidambar*, maidenhair tree, *Gingko*, and tulip tree, *Liriodendron*. Magnolias evolved even before bees and their flowers are extremely robust in order to cope with the weight of their beetle pollinators.

Kew botanists and horticulturists continue to enrich botanical collections worldwide, replenishing stock with seed gathered under license from populations in the wild.

The Mighty Oak and its Relatives

The oak family, Fagaceae, provides Kew Gardens with some of its most impressive trees, including the largest, oldest and – perhaps – most curious.

The chestnut-leaved oak, *Quercus castaneifolia* **M6** is the largest tree in the Gardens. Coming from the Caucasus and Iran, no finer example of this species is known in the world. It is widely acknowledged to be one of the biggest trees in Britain, measuring over 30m tall and 30m wide. Planted in 1846, it has reached its great size in just under 170 years, making it the fastest growing tree in the Gardens.

The chestnut-leaved oak is so called because its leaves are likened to those of Kew's oldest planted tree, a sweet chestnut, *Castanea sativa*. The gnarled and twisting specimen near the Mediterranean Garden **H6** exhibits features that are rarely seen in its native habitat, the dense chestnut forests of southern Europe. The watchful care of generations of Kew arboriculturists has given this and other nearby trees the space and long life to develop such intriguing character. Also belonging to the oak family, the sweet chestnut is thought to have been introduced to Britain by the Romans and it is an important element of ancient coppiced woodland in southern England.

A rival for the title of oldest tree also pre-dates Princess Augusta's botanical garden. Near the head of the Lake **H5** and not far from a line of old sweet chestnuts, stands an immense English oak, *Quercus robur*. It may have been planted over 285 years ago when the famous landscape designer Charles Bridgeman worked on this area when it was part of the Richmond estate belonging to George II and Queen Caroline (see Changing Landscape p118).

The title of 'most curious' tree at Kew may go to another member of the oak family – the weeping beech, *Fagus sylvatica* 'Pendula', which was planted in 1846 between the Broad Walk and the Ice House **M7** . Where its sinuous boughs have dived back to the ground, they have put out roots and new trunks have sprung up. Over time, repeated layering has created a vast complex of smooth grey limbs, only revealed when its magnificent display of autumn leaves is shed and the low light of the winter sun penetrates its shady cavern.

The Lockerbie oak (*Quercus robur*) by the Lake.

It is no accident that so many oaks are planted in the Gardens. There are around 600 species of oak worldwide and they are revered for their strong timber, long life and ecological value. Kew's oak collection ranges along the riverside, and many huge oaks from North America, such as the northern red oak, *Quercus rubra*, can be found as specimen plantings, setting the Gardens ablaze with their spectacular red leaves in autumn.

In one summer a large oak may use up to 40,000 litres of water, releasing it into the atmosphere through 700,000 leaves. It produces enough oxygen in a year for ten people.

Conifers

Conifers make powerful statements. At Kew you can find them massed in the Pinetum where the wealth of form and colour within this ancient and enigmatic group of plants becomes apparent.

Only three conifers are native to Britain – the Scots pine, *Pinus sylvestris*, yew, *Taxus baccata*, and juniper, *Juniperus communis*. As travel became more commonplace in the 18th century, owners of the great estates craved the drama brought to the landscape by the cypress of Italy, the cedar of the Levant and the white pine of North America. The early 19th century brought great numbers of conifer species to the UK and one conifer from Kew's original five acre arboretum survives – the Corsican pine, *Pinus nigra*, planted in 1814 and still standing near Elizabeth Gate.

Sir Joseph Hooker was fascinated by conifers, both for their beauty and their economic value. In the 1860s he developed the Pinetum **C4** on the south side of Kew's Lake. Wandering through the groves, the extraordinary diversity of cone colours and shapes can be seen amongst the dark needle-like foliage, from the spiky spiralled candles of the pines to the bright open 'berries' of the yew. Conifers are pollinated by the wind, their lightweight pollen grains borne through the air to fertilise female cones of the same species, which then produce seed.

Growing in the Gardens are several grand specimens of stone pine, *Pinus pinea,* with its distinctive umbrella-like structure. Highly valued for its edible seeds known as 'pine nuts', it has been cultivated in Europe for almost 2,000 years.

Conifers appeared over 300 million years ago – indeed, their fossilised remains help scientists trace the geological history of the Earth. Over the millennia they have struggled to compete with the flowering trees that evolved later. While there are believed to be between 60,000 and 80,000 tree species globally, there are just 615 species of conifer, 200 of which occur in tropical forests and savannahs.

Over a third of conifers are now under threat, through habitat destruction and over-exploitation. World knowledge of this fascinating plant group has been dramatically improved with the work of Kew experts.

One of the most atmospheric places in the Arboretum is the towering Redwood Grove **C4** , planted in the 1860s, where giant redwoods,

Sequoiadendron giganteum, stand in solemn majesty with coastal redwoods, *Sequoia sempervirens*. Among them is Kew's tallest tree, a coastal redwood, which at nearly 40m is the height of a 13-storey building. A tree of the same species growing in Redwood National Park, California, is believed to be the world's tallest living organism, reaching nearly three times higher than Kew's giant, at 115m high.

The atmospheric Redwood Grove, off Cedar Vista, is home to Kew's tallest tree.

The Living Fossils

The longevity, beauty and symbolic power of two ancient tree species has assured their future through cultivation.

THE NOBLE PINE

In 1994 the discovery of a conifer in a remote canyon in south-east Australia sent shock-waves through the botanical world. Similar trees were known from the fossil record, but were thought to have become extinct millions of years ago. It became clear that this was a species new to science. It was named *Wollemia nobilis*, the Wollemi pine, after the region where it was found and the man who found it, David Noble. The remaining natural populations – around 100 mature trees in all – grow in deep sandstone gorges, the tallest being 40m high and perhaps 1,000 years old.

Kew has helped to secure the future of this species by conducting hardiness trials over several years, proving that young trees coped well with the UK climate. The Wollemi pine has become prized by gardeners, and propagated plants are now sold around the world, raising money to help conserve the wild populations. Kew's Wollemi pines can be found on the Orangery Lawn **O6** , in a group near the Treetop Walkway **F5** , and on the Lake islands **F3** .

Unlike their close relative the monkey puzzle, *Araucaria araucana*, single trees develop numerous trunks with characteristic knobbly dark-brown bark that resembles bubbling chocolate. The tree is evergreen, but its pendulous frond-like foliage changes from fresh apple-green to bluish as it ages and in colder months is tinged with bronze.

Above: Wollemi pine branch with round, female seed cone, Right: the leaves become a yellowish-dark green colour with age and the rough bark resembles bubbling chocolate.

The ancient ginkgo at Kew was originally planted next to Princess Augusta's 'Great Stove' glasshouse – now long gone. It is a Champion Tree – meaning it is one of the most notable in this country.

TRUE SURVIVOR

Another tree that time forgot is the maidenhair tree, *Ginkgo biloba*. It is the only surviving member of an ancient group of plants that were widespread at the same time as the dinosaurs, 180–200 million years ago. But *Ginkgo* barely survived the ice ages in scattered pockets of eastern and southern China.

Like the Wollemi pine, its best hope for a future was mankind. Thankfully, early modern people found medicinal and cultural uses for it. Its distinctive leaves and great longevity have brought it a powerful symbolic role and many ancient specimens – up to 3,500 years old – grace the grounds of Buddhist temples.

Though endangered in the wild, the *Ginkgo* has re-colonised the world through human activity. It was brought to Europe from Japan around 1730 and arrived in North America soon after. Its resistance to disease and pollution make it a valued addition to urban landscapes and it is now one of the most common street trees in New York City. The extracts of its leaves are being investigated in biomedical research, informed by Kew's detailed studies of its traditional medicinal uses.

Kew has many examples of this beautiful tree. Our oldest *Ginkgo biloba* is thought to have been planted around 1762 in the original botanic garden of Princess Augusta **07** .

Natural Areas at Kew

While Kew is best known for its work with plants from across the globe, there is a quiet enclave devoted exclusively to habitats of south-east England

The Natural Areas **B2**, at the south-western end of the Arboretum, are a mosaic of meadow and woodland beside the River Thames. They are managed to encourage native flora and fauna. A network of sinuous paths leads the visitor into a wild, romantic place quite different from the stately vistas and designed landscapes that surround it.

These 15 hectares of sylvan beauty were given to Kew by Queen Victoria in 1898, on the condition that they remained in their natural state. The gift included Queen Charlotte's Cottage, built in the 1760s in a woodland glade (see Changing Landscape p118).

At that time it was thought that conservation could be achieved simply by fencing off areas and leaving them to natural processes without any interference. The intervening century has seen a resurgence of traditional cycles of woodland and meadow management, as their importance to wildlife and diversity has been rediscovered. Kew has embraced this approach: the removal of non-native trees and shrubs, judicious mowing, the introduction of coppicing and the protection of small wetland areas all play a part in the ongoing development of this precious retreat, while standing dead trees are retained as habitats for invertebrates, fungi, bats and birds.

The tree community is made up of a mix of mature native broadleaf species such as oak, *Quercus robur*, hornbeam, *Carpinus betulus*, beech, *Fagus sylvatica*, alder, *Alnus glutinosa*, and ash, *Fraxinus excelsior*, with naturalised species of horse chestnut, *Aesculus hippocastanum,* and sweet chestnut, *Castanea sativa*. The understorey has been dominated by yew, *Taxus baccata*, and holly, *Ilex aquifolium*, but these evergreens are being thinned in favour of plants that cast less shade, such as mountain ash, *Sorbus aucuparia*, hawthorn, *Crataegus monogyna*, and spindle, *Euonymus europaea*. You can see many of these species and learn more about them on the Woodland Walk **B1** – a boardwalk trail through a previously hidden glade.

Apart from providing beautiful and peaceful places, woodlands have many benefits including making us more resilient to climate change, conserving biodiversity, contributing to a sustainable green economy and improving our health and wellbeing. Experts from Kew's Millennium Seed Bank are working with the Forestry Commission on the UK National Tree Seed Project, collecting seeds of all the UK's native tree and shrub species. The collection will support conservation and research to help achieve national goals of increasing woodland cover across the UK.

Flowers in the Woodland garden.

MANAGING WOODLAND FOR WILDLIFE

The traditional management of Britain's woodlands creates valuable wildlife habitats. Historically they have also been put to many uses by people. The sessile oak, *Quercus petraea*, was used to build ships as well as houses, while understorey shrubs were utilised for fencing, tool-making and fuel. Coppicing, a sustainable method of harvesting, was once common in Britain, and Kew has reintroduced the practice into the Natural Areas.

Hazel, *Corylus avellana*, sweet chestnut, *Castanea sativa*, and lime, *Tilia*, are left to grow as multi-stemmed shrubs for a few years and then their wood is harvested by cutting right back to their bases. All three species regenerate quickly, producing long straight stems that can be harvested again within a few years. The wood is used in the Gardens for plant supports and fencing. Coppicing is carried out on a cyclical basis, bringing light into the glades, encouraging flowers like snowdrops, wild garlic and bluebells.

Rides, glades, coppice and wetlands encourage diversity in the Natural Areas, which ring with the song of birds like blackcap, dunnock, garden warbler, nuthatch and long-tailed tit. Greenfinches, chaffinches and treecreepers add to the soundscape, while sparrowhawks hunt silently through the canopy and kestrels hover over grassy margins, looking for voles. Foxes and weasels can occasionally be seen, as can smaller mammals such as wood mouse and common shrew.

While decaying ancient trees, standing dead wood and falling timber have in former times been seen as symbols of neglect, they are now recognised as one of the Natural Areas' most important habitats. Up to a third of the woodland invertebrates depend on dead wood at some stage in their lifecycle, in turn supporting creatures further up the food chain like the great spotted woodpecker, which can be heard hammering at standing dead wood in its search for food. The high volume of dead wood in the area has also encouraged a rich fungal diversity that is not so evident in the rest of the Gardens. Kew's Stag Beetle Loggery **B3** provides a safe haven for a globally endangered creature that depends on dead wood.

The stag beetle is often seen at Kew in May and June. Its fearsome-looking antlers are actually jaws, but it is harmless, helping to return minerals from dead plants back to the soil. The larvae mature for up to seven years in rotting wood before turning into beetles.

Wetland is a further element in this wildlife tapestry. In the wildlife pond waterboatman, mayfly and damselfly are joined by amphibians – the common frog, common toad and the common or smooth newt. Visitors can watch the wildlife and learn more about it on the Woodland Walk trail **B1** .

Long-grass areas bring a naturalistic beauty here and across the Arboretum. Careful management is required to maintain a balance of plant species. Grasslands and meadows are valuable for invertebrates and these features are managed to maximise biodiversity. Grasshoppers and crickets lay eggs at the base of the grassland sward in late summer, so grass-cutting takes place as late in the year as possible and the clippings are removed. Kew ecologists are also helping to conserve acid grassland, an endangered native habitat which is important not only for rare plant species but also for the wide range of insects, reptiles and birds it supports.

Art

Since the cave paintings of prehistoric times people have used art to represent nature, but over the last 400 years the science of botany has pushed the artist to new aesthetic and technical heights. Kew's world-leading collection of botanical art and illustration includes works by the great masters of the 18th century through to the celebrated artists who collaborate with our scientists today.

Throughout the Gardens, sculptures and statues are incorporated into the landscape and changing exhibitions seek to explore and enhance our profound connections with nature. Our rich collections of art, archives and artefacts show how people have made those connections with the wider world, through exploration, observation and craft.

Art and Artefacts at Kew

Kew is home to one of the world's greatest collections of botanical illustration, as well as a fine collection of portraits of people associated with botany and horticulture and over 250 objets d'art, ranging from sculptures to scientific instruments.

Alongside such art, the craft skills of cultures around the world are captured in Kew's Economic Botany Collection, the largest of its kind, with over 100,000 objects illustrating human use of plants, from ancient Egypt to the current day. Founded in 1847 as the Museum of Economic Botany, it displayed all manner of plant materials, from food and medicine to textiles, wood and rubber. The museum displays formed a bridge between producers of raw materials worldwide, and manufacturers in Britain. In the 21st century the Collection has taken on a new role, as a record of Kew's history, as an inspiration to artists and makers, and as a resource for the source communities from whom these specimens came in the first place.

Often very beautiful in their own right are the letters, sketchbooks and field notes of botanists, gardeners and travellers held in Kew's archives. This rich historical resource – over 7 million sheets of paper – charts the discovery, study, transfer and use of the world's plants and fungi.

Kew's archives include extraordinary jewels such as Sir Joseph Hooker's botanical drawings from his travels in the Antarctic and India, joined by his letters and notebooks, embellished with sketches of plants and landscape.

Also behind the scenes is Kew Library's collection of illustrated books, which includes fine volumes from the private library of Kew's first official director Sir William Hooker, such as Besler's *Hortus Eystettensis* of 1613, and other highly illustrated books including rarities like Jacquin's *Selectarum stirpium Americanarum historia* and John Sibthorp's handcoloured ten-volume *Flora Graeca,* produced between 1806 and 1840, as well as many beautiful recently published items.

So how do our visitors explore these treasures? Our collections of botanical art and artefacts are regularly drawn upon for exhibitions in the Shirley Sherwood Gallery of Botanical Art, while the unique Marianne North Gallery encompasses Victorian history at its most decorative and eccentric. Statues and sculpture can be found throughout the Gardens, bringing another dimension to the changing landscape. The permanent collection is joined by many installations and works on loan for specific festivals, such as exhibitions of world-famous artists including Henry Moore, and Dale Chihuly.

The Economic Botany Collection at Kew illustrates the extent of human use of plants around the world, from food, medicine and utensils, to social activities and clothing. The collection forms an important bridge between biological and cultural diversity, valuable for the study of plant uses past, present and future.

You can book an appointment with Kew's archives team to view the original archive items in the public reading room. Items from the Library, one of the most important botanical references sources in the world, are also made available in the reading room. You can learn more about Kew's collections through Kew's website, www.kew.org.

Treasures of Botanical Art

The Shirley Sherwood Gallery of Botanical Art **F9** opened in April 2008, as the first and only purpose-designed botanical art gallery in the world. Showcasing original botanical paintings from the Shirley Sherwood Collection together with curated shows drawn from Kew's extensive and important collections, this elegant gallery has gained a worldwide reputation for exhibitions dedicated to promoting the genre of botanical art.

By 2022, the gallery had welcomed over a million visitors and staged more than 50 exhibitions. In recent years it has broadened the scope of its programme by featuring some of the world's most exciting contemporary artists who work with, or are inspired by Kew's work and collections, highlighting themes such as biodiversity loss, food security and climate change, in innovative ways. Two major exhibitions are presented each year across five gallery spaces, from thematic group exhibitions to in-depth solo presentations. Each is accompanied by curated exhibition from the Shirley Sherwood collection inspired by the main themes.

Botanical illustration of silver-leaf wheel pincushion (*Leucospermum formosum*) published in *Curtis's Botanical Magazine*, 1815.

BOTANICAL ART

The primary purpose of botanical art is to reproduce in scientific accuracy every detail of the plant portrayed. It evolved as a genre in the age of scientific discovery in the 18th century when botanists sought to identify and classify plants new to western science. Images were needed to accompany publications, thus the practice of artists working in tandem with botanists began. It is a practice still used today.

Contemporary botanical artists often follow the pictorial conventions of the traditional plant portrait that were established in the golden age of botanical art. These include working in watercolour on paper as their primary medium, setting the single plant against a white background and including the full life cycle of the plant from bud to fruit, often with details of bisected seeds and reproductive parts.

In recent years contemporary practitioners have explored how far the conventions of traditional botanical art can be stretched, with innovative compositions, different media and scale whilst remaining true to the strict rules of scientific accuracy.

The Shirley Sherwood Gallery of Botanical Art.

THE SHIRLEY SHERWOOD COLLECTION

Interested in plants and art since childhood, Dr Shirley Sherwood started her Collection in 1990. It now comprises over a thousand paintings and drawings, representing the work of over 250 contemporary artists from 30 countries around the world. Alongside the curation of exhibitions of works from her collection for this gallery, she has exhibited in many other prestigious locations including the Smithsonian Institution in Washington, the Ashmolean Museum in Oxford and the Real Jardín Botánico in Madrid.

*Ipomoea and Vavangue with
Mahe Harbour in the distance,*
by Marianne North, 1833.

Marianne North Gallery

The Marianne North Gallery F9 was purpose built in 1882 to house an exuberant display of botanical and landscape paintings by an intrepid and unconventional Victorian traveller.

Marianne North was 40 when she embarked on her first solo expedition in 1871, determined to 'paint the peculiar vegetation' of other lands. Over the next 13 years she travelled widely to destinations including North and South America, Japan, India, Australia and South Africa, capturing over 900 species in her own distinctive style. With minimal formal training, she worked using a palette of bold assertive colours, sketching rapidly in pen and ink on heavy paper, and painting with oils straight from the tube. In contrast to the scientific accuracy of botanical illustration, Marianne showed her subjects in their natural settings, sometimes with an associated insect, bird or other small creature. Her vast collection of 848 paintings constitutes a snapshot in time of tropical and temperate habitats from across the globe.

In 1879, Marianne wrote to Sir Joseph Hooker offering Kew her collection and a gallery to house it. Her architect friend James Fergusson designed the Gallery which, with its veranda, echoes her attachment to India, while satisfying his own ideas on lighting with large clerestory windows high above the paintings.

Marianne took charge of the hanging, arranging her pictures in geographical sequence over a dado of 246 strips of different timbers. An extensive two-year project has now restored the Marianne North Gallery and its paintings to their original splendour and a linking gallery joins the Grade II listed building to the Shirley Sherwood Gallery of Botanical Art. In the exhibition room, visitors can watch a short film about the artist and use interactive screens to view a set of 'then and now' images, which show how some of the landscapes visited and painted by Marianne North have changed.

Apart from this monument to an extraordinary personality – and to the world view that prevailed in Victorian Britain – Kew's collections are enriched by associated material from Marianne North. Within the photographic collection there are several original photos of her taken by the famous photographer Julia Margaret Cameron (1815–1879), while Kew's archive holds letters from Marianne to various friends and acquaintances about both her travel experiences and the creation of the North Gallery at Kew Gardens, some including sketches.

Right: The 848 paintings are arranged geographically filling every inch of space in the purpose-built Victorian gallery. They were hung by Marianne North herself (above), seen painting in South Africa.

Architecture

Kew Gardens began as two great royal estates and went on to become an economic powerhouse of empire. Now one of the world's leading centres for plant and fungi research, it is also a heritage site of global importance. Its buildings come from every stage of its history, some ornamental, others designed to provide its valuable collections with the environments they need.

In every case, the architects have employed the latest ideas and technology to ensure that their buildings are both fit for purpose and beautiful to behold.

Kew has 38 listed buildings including glasshouses, gateways and follies. Alongside these sit the contemporary designs of the modern era. The development of architectural design over time is reflected in these buildings.

The Temperate House

Buildings at Kew

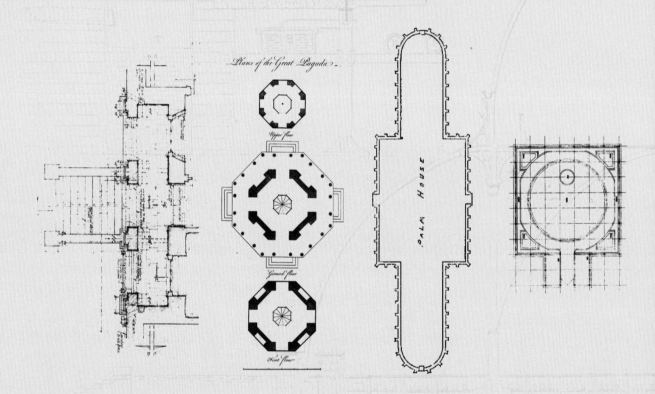

Plans of the Great Pagoda

Kew Palace

1631. Built for the rich Flemish merchant Samuel Fortrey, this palace became a country retreat for George III and his family.

Pagoda

1762. A fashionable Chinoiserie folly designed by Sir Willam Chambers to grace the pleasure gardens of Princess Augusta.

Palm House

1848. The Palm House used the latest innovations in ship-building and glazing to create a palace for rainforest plants.

Waterlily House

1852. Once the widest single-span glasshouse in the world, built to house tropical waterlilies and climbing plants.

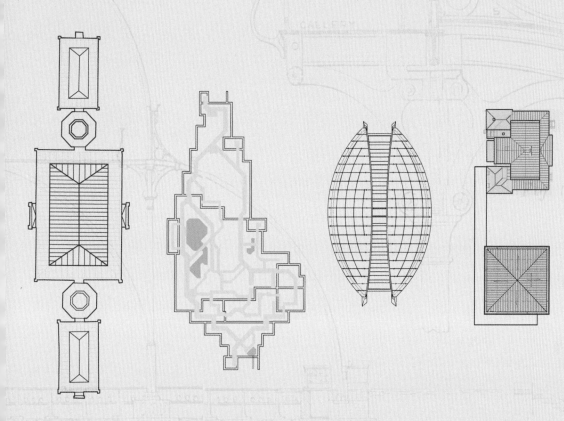

Temperate House

1862–1899. The Temperate House was built over several decades, with the central block and octagons completed first to house frost-tender plants.

Princess of Wales Conservatory

1987. Ten landscaped zones to house plants from cool deserts to tropical mountains, cloud forests and rainforests.

Davies Alpine House

2006. This ingenious new alpine house is designed like a cooling tower for plants from the world's mountain regions.

Shirley Sherwood Gallery of Botanical Art and Marianne North Gallery

2008 and 1882. These two purpose-built galleries hold magnificent collections of botanical illustration and art.

Royal Treasures

The first century of Kew Gardens was intimately bound up with the Royal Family. This peaceful, leafy estate along the River Thames was a retreat from the demands of court and the filth and chaos of London.

George II and his descendants built palaces, follies and temples to adorn the landscape and to provide accommodation fit for kings. But while Kew Palace **04** is the oldest building in the Gardens and the last surviving royal residence, its modest interiors encapsulate Georgian taste and style, rather than palatial opulence. Built in 1631 for a rich Flemish merchant called Samuel Fortrey, it was originally known as the Dutch House and the first royal residents were the three elder daughters of George II.

Their brother Frederick, Prince of Wales, lived opposite them in a large Palladian palace called the White House – since demolished – and it was here that his son, the future King George III, grew up. At Kew Palace he received the extensive education that was to profoundly influence his reign, a period in which ideas of science, art and manufacturing blossomed. Even when he became King, George disliked court life, retreating to his country estates to explore the latest ideas in agriculture and architecture. He and his wife, Queen Charlotte, spent happy summers at Kew Palace with their 15 children and it was an important refuge during his infamous episodes of 'madness'. After Queen Charlotte died in 1818, Kew Palace was shut up.

It was acquired by Kew in 1898 and opened to the public for the first time. Today it is in the trust of Historic Royal Palaces.

Close to Kew Palace, the Georgian Royal Kitchens remain miraculously preserved, 200 years after they were last used. Designed by one of the greatest architects of his day, William Kent, their dimensions give us an idea of the grand scale of the White House. The great kitchen is an impressive double-height room, complete with roasting range, charcoal grill and pastry oven, while in four adjacent rooms foodstuffs were stored, and pots and pans scoured. Upstairs is the office of the clerk, responsible for feeding the enormous royal household. He held under lock and key the expensive spices, sugar, cinnamon and other luxuries.

Kew Palace.

Kew's handsome classical Nash Conservatory is the oldest of Kew's 19th century glasshouses. Originally one of four pavilions designed by John Nash for the gardens at Buckingham Palace, it was moved to Kew by King William IV in 1836.

Following a major restoration project, in the summer months you can climb the heights of the Great Pagoda and marvel at spectacular views across London.

CLASSICAL PROPORTION

The Orangery **05** is a fine classical building constructed in 1761 and designed by Sir William Chambers to house citrus trees and other ornamentals. However, even with the addition of two side windows in 1842 the light levels were never high enough for plants to thrive inside.

AUDACIOUS FOLLY

The Great Pagoda **A8** was completed in 1762 as a gift for Princess Augusta, the founder of the botanic gardens at Kew.

It was one of several Chinese buildings here, including a bridge and an aviary, designed by royal architect Sir William Chambers at a time when Chinoiserie was the height of fashion. The Pagoda was originally flanked by a Moorish Alhambra and a Turkish Mosque – follies that helped bring the architecture of the world to Kew. As symbols of fashion, wealth and power, these garden ornaments were the subject of much public comment. The Pagoda was judged 'a whimsical object', 'the puerile effort of an overgrown boy' and 'a masterpiece of art'.

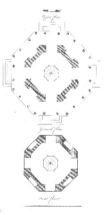

During the construction of the Pagoda, the politician Horace Walpole, who lived in nearby Twickenham, wrote 'We begin to perceive the tower of Kew from Montpelier Row; in a fortnight you will be able to see it in Yorkshire.' The ten-storey octagonal structure is nearly 50 m high (163 ft) and the uppermost level is reached by a staircase of 253 steps. Contemporary commentators wondered that such a tall building could remain standing but its tapering design gives it great stability, each successive floor being 30 cm (1 ft) less in both diameter and height than the one below it. Throughout the 20th century the Great Pagoda was painted a bright postbox-red, but Chambers' original colour scheme was much more colourful.

Historic Royal Palaces has restored this glorious folly and their detailed paint analysis has shown that the window frames and balustrades were once a warm stony white, while the window casements and the top roof were Venetian red. The undersides of the ten roofs were striped in green and white, which change in ratio to give the building an impression of being taller than it is. A return to Chambers' 18th-century scheme brings historic integrity back to this spectacular building, but, best of all, it has also brought back the dragons. The Pagoda was designed and built with 80 crouching dragons perching on the hips of the roofs, but little was known about them as they disappeared in 1784.

After much research, they have been recreated and painted in original bright colours of 'copper verdigris', red, blue and gold, adding wonderful drama to this corner of the Gardens.

Visit the Great Pagoda **A8** and take in the stunning views from the top while learning more about Kew and the Georgians. Nearby you'll find a new Chinese landscape full of beautiful plants of Chinese origin and the stories of how they came to our gardens.

Palaces of Glass

One of the most important Victorian structures still in existence, the Palm House has stood at the heart of Kew Gardens for 170 years.

Shortly after the Gardens passed from the Crown to the nation the Palm House **K7** was built to meet the urgent need to house the palms and other tropical trees that were being dispatched to Kew from around the world. It was the hub of Nesfield's landscape design, which unified the historically distinct estates, with vistas radiating from its west doors to the far reaches of the site.

THE PALM HOUSE

The Palm House was built using the latest technological innovations. It was constructed between 1844 and 1848 by engineer Richard Turner to Decimus Burton's design. The elegant curvilinear structure, with its large central hall unimpeded by pillars, was only made possible by developments in shipbuilding and glazing. It was heated by a system of pipes fed by 12 boilers, circulating warm air through an iron grating floor. An underground tunnel was used to deliver fuel to the furnaces and carried the smoke away to a chimney, disguised as an Italian Romanesque campanile, which can still be seen towering over Victoria Gate.

The Palm House measures 110 m (363 ft) long, 30 m (100 ft) wide and 19 m (63 ft) high. In spite of the fact that it was originally clad in 16,000 panes of green-tinted glass, it eventually proved to be a horticultural success. The first palms, cycads and trees in the Palm House were planted in large teak tubs or clay pots and within five years Turner's elegant spiral staircases and wrought iron balustrades were smothered by climbers. In 1860, two large central beds were dug and the tallest palms planted in them. Freed from their pots the trees thrived, several species flowering for the first time outside the tropics.

The Palm House was first restored between 1955 and 1957 when its glazing bars were cleaned and the entire house re-glazed. A second more comprehensive overhaul took place between 1984 and 1988. The Grade I listed-building was emptied, completely dismantled, restored and rebuilt. Ten miles of replica glazing bars made of stainless steel were put in place to hold new panes of toughened safety glass. The restoration took as long to finish as the glasshouse first took to build.

The Davies Alpine House and the
Princess of Wales Conservatory.

In the Temperate House you can discover stories of rare and threatened plants and Kew's work in the temperate regions of the world to study and conserve plant diversity for the benefit of everyone.

THE TEMPERATE HOUSE

Once the largest glasshouse in the world and now the world's largest surviving Victorian glass structure, the Temperate House **E7** underwent an ambitious five-year architectural restoration, reopening in 2018.

Another spectacular Decimus Burton design, the Temperate House is twice the size of the Palm House at 4,880 square metres. It was built over several decades (1862–1899) to house frost-tender plants from the sub-tropical and warm temperate regions of the world. Burton designed the planting beds so that species could be displayed by geographical region and this is still the case today.

In the central section of the house you'll find plants from Central and South America, New Zealand, Australia and from islands around the world including those on which Kew has active conservation programmes including St Helena. The north wing holds plants from Asia, while in the south wing you'll discover the flora of South Africa. In total the Temperate House is now home to over 10,000 plants of 1,500 species – an extremely important scientific collection that is both vital for conservation and beautiful to visit.

The restoration of this complex and ornate glasshouse has been done to exacting heritage standards and included repairing much of the elaborate stonework, urns, statuary, window frames and ironwork. Extensive new leading had to be completed and 15,000 panes of glass were replaced. New walkways throughout the house afford visitors better views of the plants and the stunning new water features. The octagons are now places to enjoy changing educational and horticultural displays. Take time to explore and delight in the restored architecture and the many special and useful plants on display here. Make sure you talk to the explainers, who are on hand to share intriguing plant stories and can point out the highlights of the collections, including the Wood's cycad, *Encephalartos woodii*, St Helena olive, *Nesiota elliptica*, marsh heath, *Erica verticillata*, *Banksia brownii* and *Dombeya mauritiana* to name but a few.

THE WATERLILY HOUSE

Kew's Waterlily House **K6** opened in 1852. At the time it was the widest single-span glasshouse in the world, purpose-built to house the natural wonder of the age, the giant waterlily, *Victoria amazonica*. Today, you can see its close relative *Victoria cruziana* here. *V. amazonica* is grown in the Princess of Wales Conservatory, alongside the newly described species *Victoria boliviana*. All three put on glorious displays every summer, along with *Nymphaea* waterlilies and sacred lotus, *Nelumbo nucifera*.

Contemporary Icons

The architects of Kew's extraordinary 19th-century glasshouses created structures that were not only beautiful but purpose-built for the greatest plant collection in the world. Two award-winning glasshouses at the north end of the Gardens extend this fine tradition into the modern age.

PRINCESS OF WALES CONSERVATORY

The Princess of Wales Conservatory **N7** was opened in 1987 by Diana, Princess of Wales, and is named after her predecessor Princess Augusta, the founder of the botanic gardens at Kew. Its dramatic angled gables and its surrounding landscaped beds are a sleek rejoinder to the curves of the Palm House and the classical ornament of the Temperate House. Designed by Gordon Wilson to achieve the highest possible energy efficiency, its stepped glass roof is an extremely effective collector of solar energy and much of the building lies below ground level, which insulates it against heat loss. Inside, paths, steps, bridges and tunnels lead the visitor through ten landscaped zones, their temperature and humidity carefully controlled to recreate a range of environments, from tropical desert to rainforest (see p18).

DAVIES ALPINE HOUSE

The most recent glasshouse built at Kew is also the smallest. The glass and steel arch of the Davies Alpine House **08** rises lightly over the northern sandstone terraces of the Rock Garden. It is designed to keep plants dry in winter (when they would naturally lie under a blanket of snow) and provide the cool summer conditions of their mountain habitats, without using energy-intensive air-conditioning and wind pumps. To counteract the build up of heat that always occurs under glass, the distinctive structure acts like a chimney; as warm air escapes through vents in the roof, cool air is drawn in at the base. Meanwhile, a fan blows air through a concrete labyrinth beneath the ground. The air cools on its convoluted journey and flows over the plants from a series of funnels within the rocky landscape. The laminated glass has an extremely low iron content, allowing 90% of daylight through. Alpines from high altitudes thrive in the resulting intense sunlight. However, in extremely hot periods the temperature in the house is regulated by shading provided by four beautiful fan-shaped sails. Designed by Wilkinson-Eyre Associates, the Davies Alpine House was awarded a prestigious RIBA award for architecture when it opened in 2006.

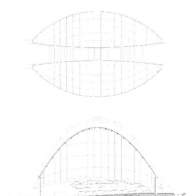

TREETOP WALKWAY

While Kew's glasshouses and galleries focus on their contents, another striking contemporary structure offers thrilling perspectives on the great outdoors. At 18 m (59 ft) above the ground, Kew's Treetop Walkway **F5** gives visitors a rare insight into the complex ecosystem of the forest canopy, a world teeming with birds and insects, lichens and fungi. The 200 m (656 ft) long walkway was opened in 2008, designed by Marks Barfield Architects, who also designed the London Eye, and won numerous awards in 2009 including the RIBA Award, the Structural Steel Design Award and the Civic Trust Award. Made from over 400 tonnes of weathered steel, the 12 walkway trusses connect ten 'node' platforms, including one large enough to provide a lofty, open air classroom for school groups. Far-reaching views across the Gardens and over the London skyline show the vital part trees play in the landscape, not only contributing to its beauty but providing habitats for urban wildlife and acting as natural air conditioners (just a 10% increase in urban greenery can cut the temperature in a city by up to four degrees centigrade in summer).

LAKE CROSSING

Opened in 2006 the Lake Crossing **G4** is the first ever bridge across the Lake, designed by the architect John Pawson. Winner of the 2008 Stephen Lawrence Prize, the striking black granite walkway guides visitors low over the water along a curving path that mimics the Lake's rounded banks. Its walls are a series of vertical, flat bronze posts. On approaching the Crossing these give the appearance of forming a solid wall, but when viewed sideways on they appear almost invisible. This is akin to the ways in which water can appear both solid and fluid.

SHIRLEY SHERWOOD GALLERY OF BOTANICAL ART

The Shirley Sherwood Gallery of Botanical Art **F9** is the first public gallery dedicated to botanical art, displaying treasures from Kew's collections (see p84–5) alongside works from the collection of Dr Shirley Sherwood and partner institutions. Award-winning architects Walters and Cohen employed the latest technologies to provide controlled conditions to keep the artworks in perfect condition, but with minimal environmental impact. In 2023, the Gallery celebrated its fifteenth anniversary.

Left: The design and trusses of the Treetop Walkway are inspired by the Fibonacci sequence, a number series found frequently in natural growth patterns.

THE HIVE

The Hive N6 is a unique and critically acclaimed structure, inspired by scientific research into the health of bees. Designed by UK-based artist Wolfgang Buttress, it was originally created as the centrepiece of the UK Pavilion at the 2015 World Expo in Milan.

The Hive is an open-air structure standing at 17 m (56 ft) tall and weighing in at 40 tonnes. It is made from around 170,000 individual components that together create an intriguing lattice or honeycomblike effect. Inside you'll find around 1,000 LED lights that glow and fade as a unique soundscape hums and buzzes around you. The Hive encapsulates the story of honeybees and their important role in pollinating crops and other plants.

The multi-sensory elements of light and sound within the Hive are in fact responding to the real-time activity of bees in a beehive behind the scenes at Kew. Their intensity changes as the energy levels in the real beehive surge, giving visitors an insight into life inside a bee colony.

Around the Hive you'll find a meadow with wildflower species beneficial to bees, including selfheal, ox-eye daisies and ragged robin, which flower over the summer and autumn.

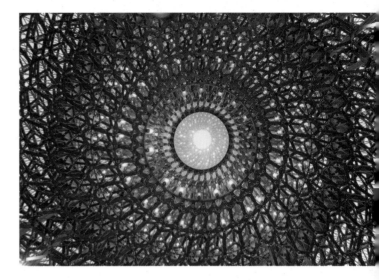

The Hive moved to Kew in 2016 and took four months to reconstruct. Now one of the most unique features in the Gardens it won a Landscape Institute Award for its immersive impact. A beacon of contemporary art, it continues to stand tall as a striking symbol of the challenges facing bees today.

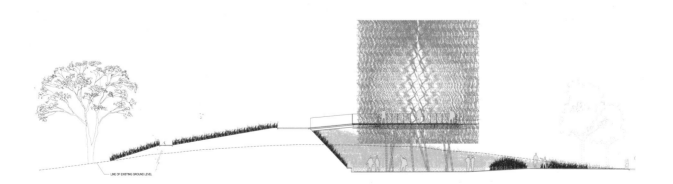

LINE OF EXISTING GROUND LEVEL

Changing Landscape

The landscape of Kew Gardens is as significant as its plant collections and its architecture. The vistas and glades represent ideas about nature and science as they have developed over 260 years.

Lakes and lawns frame Kew's iconic buildings and the great displays of trees and flowers paint complex ever-changing panoramas through which the visitor can explore this tranquil, historic place.

Living History

Kew's complex history is imprinted upon its landscape. Its contours and plant collections chart three centuries of development, from royal pleasure ground and kitchen garden to people's park, from pastoral idyll through the frenzied collection of plants and the industrial exploitation of natural resources, to a national reference collection informing conservation, medical and genetic research.

Now encompassing 132 hectares (326 acres), the site is bound by the River Thames to the west, the beautiful eastern wall along Kew Road, the elegant Kew Green to the north, and Richmond's leafy deer park to the south. At every point, the visitor is beckoned by views, ornaments, buildings and extraordinary plants, each with a fascinating story. These multiple layers of history, science, culture and personality are celebrated in Kew's status as a UNESCO World Heritage Site.

TREND-SETTING ROYAL ESTATES

The Gardens were born of the ambitions of two powerful 18th-century women who sought to present the latest fashionable thinking through their adjacent royal estates. Queen Caroline pioneered the English Landscape Garden style – with no expense spared – while her daughter-in-law, Princess Augusta, embraced Enlightenment ideas of purpose and pleasure by founding a botanic garden at the heart of her Chinoiserie-style landscape.

Queen Caroline, wife of King George II, engaged Charles Bridgeman to redesign Richmond Gardens in the mid-1720s. Formal baroque terraces and avenues were swept away and this flat expanse of floodplain, with the River Thames at its western boundary, was sculpted and planted to create an Arcadian landscape of walks and woodland, interspersed by arable fields. William Kent's Hermitage, his famous gothic Merlin's Cave and other follies enhanced this recreation of rural idyll. Lancelot 'Capability' Brown, one of the most famous landscapers in Britain's history, further developed the naturalistic character of Richmond Gardens in the first decades after George III ascended the throne. Kent's whimsical buildings and the beloved riverside terrace were replaced by sweeping lawns and artful clusters of trees. Brown's picturesque Hollow Walk of 1773 survives as the Rhododendron Dell **13** .

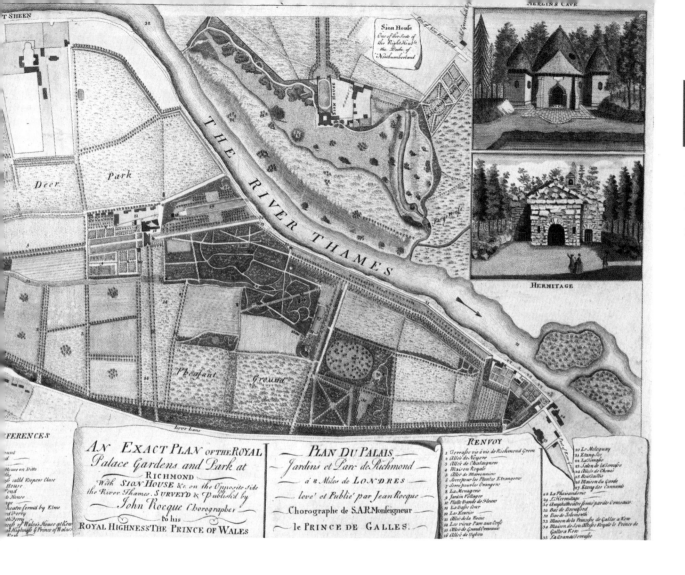

An *exact plan* of the Royal Palace Gardens and Park at Richmond, with Syon House on the opposite side of the River Thames, 1754.

Another rustic element that can still be seen in the south-western corner of today's Gardens is Queen Charlotte's Cottage **A3** . An early example of a *cottage orné*, the thatched building was used by the Royal Family in the late 18th century for rest and refreshment during walks in the gardens. Exotic animals were kept in the paddock to the rear of the cottage, including colourful pheasants and the first kangaroos to arrive in England.

John Rocque's 'exact plan' of the Royal Palace Gardens and Park at Richmond, with Syon House on the opposite side of the River Thames, 1754.

FASHION RIVALS

Princess Augusta, wife of Prince Frederick and mother of the future George III, was to promote an even more flamboyant, exotic style in her garden at Kew. Statuary, formal shrub plantings, lawns and hedged fields of the 1730s were gradually embellished by 'curious & forain' trees, a lake and mount. However, it was after Frederick's unexpected death in 1751 that work began in earnest to implement his plan for the Kew garden. Guided by the Earl of Bute, mentor to the future George III, Princess Augusta founded the botanic garden, a five acre arboretum and a physic garden. Her burgeoning tree collection – in particular, new introductions from North America – was laid out according to the new method of bi-nomial classification devised by Swedish botanist Carl Linnaeus – each plant uniquely identified by two Latin words. Within ten years, the list of Kew's plants – *Hortus Kewensis* – listed 3,400 species.

A view of the Lake and Island at Kew as seen from the Lawn by William Marlow, from William Chambers' Gardens and Buildings at Kew, 1763.

FOLLIES IN THE PLEASURE GROUNDS

At the same time, the architect Sir William Chambers began to create buildings for the Princess's pleasure grounds, among them the Great Pagoda and the Orangery. Of his many follies – ornamental architectural features of little practical use – four remain, all of them Grade II listed buildings. The Temple of Arethusa (1758) **K8** is named after a nymph attendant of Diana, Roman goddess of hunting. Today this temple is the setting for Kew's war memorials and is adorned with wreaths every 11 November. The Ruined Arch (1759) **E9** is a charming mock ruin in classical Roman style; the Temple of Bellona (1760) **I8**, with its Doric facade, is named after the Roman goddess of war; the Temple of Aeolus (1760) **M8**, named after the mythical king of storms and the four winds, was originally built of wood and once had a revolving seat. It was rebuilt in stone by Decimus Burton in 1845.

William Andrews Nesfield's design for the landscape at Kew centred around the Palm House and included the formal parterre on its east side.

ORNAMENTAL LAWNS

As European explorers charted the world, both gardens were developed to display exciting newlydiscovered plants. New technology affected not only the architecture but the landscape too. In the 18th century, Kew's pastures were grazed by livestock, but over the decades, the gradual mechanisation of grass cutting led to the long tradition of close-cropped ornamental lawns. Nineteenth-century garden design saw the lawn as the perfect foil for the major structural elements of the landscape – including the latest plant introductions from around the world, as well as terracing and parterres. In 1840, just as the idea of public parks had begun to take root, the Royal Botanic Gardens, Kew was transferred from the Crown to the nation. Under the directorship of Sir William Hooker it became a destination for public recreation and education. Its landscape was reconfigured as parkland and a fitting backdrop for monumental architecture, symbolising its new role as a driver of economic development across the British Empire.

VICTORIAN FORMALITY

William Andrews Nesfield, the leading exponent of renewed formality in gardens, unified the historically distinct estates by imposing on them a series of long vistas, which radiated out from the new Palm House, its pond and parterres. The Palm House complex was linked by the stately Broad Walk to a grand new entrance on Kew Green. Instead of an inconspicuous door in the boundary wall, visitors now entered the Gardens through Decimus Burton's ornate gateway with Jacobean-style ironwork and carved stone pillars. This gate was renamed Elizabeth Gate in 2012 to mark the Diamond Jubilee of Queen Elizabeth II.

The construction of the Lake – now transformed by a beautiful crossing G4 – and Waterlily Pond D4 added drama to the Arboretum. Popular demand for seasonal floral displays drove the development of hundreds of flower beds, incorporating the geometric style of the day, and the first Refreshment Pavilion, which opened in 1888.

SCIENTIFIC PURPOSE OF THE GARDENS

In spite of the emphasis on spectacle and public recreation, the scientific purpose of the Gardens remained at its core. A series of beds were laid out with plants grouped according to the scientific system of classification, today redesigned as the Agius Evolution Garden.

Trees and shrubs were planted taxonomically in the Arboretum, and the Pinetum extended. One of the best holly, *Ilex*, collections in Europe still remains along Holly Walk **H6** , while old excavations were exploited to create intimate oases like the Bamboo Garden **H3** and Berberis Dell **H8** .

The Hollow Walk became the Rhododendron Dell **I3** , planted with Joseph Hooker's celebrated collections from Sikkim. The Rock Garden **N8** and the first Alpine House were constructed and the Temperate House completed. By 1902, 8,000 herbaceous plants and 4,500 hardy trees and shrubs were recorded in cultivation.

There have been significant additions since the turn of the 20th century, such as the Grass Garden (see p48), the Rose Garden **J7** to the west of the Palm House and the Japanese Gateway (see p44). Cherry Walk **G7** was created and, in 1911, the Rhododendron Dell was enhanced by the Chinese rhododendrons introduced by E. H. Wilson, and new buildings such as the Princess of Wales Conservatory (see p110) and the Treetop Walkway (see p115) have been added.

The Rose Garden, to the west of the Palm House is home to 170 different species and cultivars of rose.

No-mow paths bring the twin benefits of improving accessibility and removing reliance on fossil-fuel-intensive mowing in Kew's Kitchen Garden.

EDIBLE SCIENCE: KEW'S KITCHEN GARDEN

Created on the site of the original Georgian kitchen garden which supplied produce to King George III's estate, Kew's Kitchen Garden **N8** is one of the first kitchen gardens open to the public with a sustainable focus. Biodiversity loss, food security, climate change and the cost of living are the defining issues of our time. In 2022, following a nine-month refurbishment project, the garden re-opened with solutions to these very much in mind.

To encourage the natural ecosystem, 'beneficial' and 'sacrificial' flowers and herbs are planted to support the crops, helping to avoid the use of chemicals. The annual rotation of crops creates a moving target for pests and diseases and helps to prevent nutrient depletion in the soil.

Our horticulturalists use a no-dig method to ensure good soil heath. Compost is added to the top of the beds, which suppresses weeds, and protects beneficial bacteria and helpful creatures that live just below the surface. Carbon storage is an added benefit, with no-dig cultivation conserving the carbon that is ordinarily released from the soil through digging.

This carefully tended, beautifully designed productive space, is close to the heart of many visitors, inspiring them to discover the pleasures of growing and harvesting edible plants in a more sustainable way, as well as to think about the bigger picture.

The Oak Tree Circle in the
Children's Garden.

THE CHILDREN'S GARDEN

The Children's Garden M4 , opened in 2019, excites and inspires children with a combination of beautiful planting and outdoor play. Set amongst over 100 mature trees, the site near Brentford Gate has been designed to appeal to children under 12 years old and is themed around what plants need to grow: earth, air, sun and water.

In the Earth Garden children can weave through a living bamboo tunnel, explore a jungle of palms and slide down 'worm-hole' tubes. The Air Garden includes trampolines, hammocks and a forest of 'wind flowers' that spin in the breeze. The Sun Garden is a space surrounded by flowering cherry trees and a tunnel of hoops, on which fruit trees will grow. In the Water Garden, children can pump water down rocky channels and divert the flow with sluice gates before it reaches a shallow pool.

The Pine Tree Wilderness is an area for more adventurous play, with challenging climbs, a log bridge and a 5 metre tall leaning tower. The centre point of the whole garden is an elevated walkway, the Oak Tree Circle, that gives views of the garden and takes children up to the canopy of an old oak tree.

The Role of Kew

The mission of the Royal Botanic Gardens, Kew is to understand and protect plants and fungi for the wellbeing of people and the future of all life on Earth.

Kew is an unparalleled repository of plant and fungal diversity. The 20,000 species that grow in our landscapes and glasshouses at Kew Gardens and Wakehurst, combined with more than 8 million specimens in our Herbarium, Fungarium, DNA Bank, Economic Botany Collection and Millennium Seed Bank, and our extensive Library and Archive collections, underpin vital scientific work to understand and protect plants and fungi. Nature provides species that feed, clothe and house us, keep us healthy, are a source of renewable energy, and support many livelihoods. However, unsustainable human activities, such as clearing forests and burning fossil fuels, are causing species to go extinct, changing natural climate patterns and disrupting beneficial ecosystem processes. Kew's scientific work, undertaken with our international network of partners, is critical to the well-being of all of us and the future of all life on Earth.

Conducting Scientific Research to Benefit Humanity

Kew's living and preserved collections, developed over its 260-year history, are underpinning critical research into biodiversity, conservation, and sustainable use.

Currently, two in five plants are estimated to be threatened with extinction. Using criteria such as geographic range, population size and threats, Kew scientists conduct conservation assessments and assign species to categories defined by the International Union for Conservation of Nature for its Red List of Threatened Species. These nine categories – Extinct, Extinct in the Wild, Critically Endangered, Endangered, Vulnerable, Near Threatened, Least Concern, Data Deficient and Not Evaluated – help authorities make informed decisions on which species to protect, where to create protected areas, how to allocate conservation funding and where development should and should not take place.

Of course, undertaking a conservation assessment for a particular species of plant or fungi is only possible once we know it exists. There are currently 350,000 accepted species on the World Checklist of Vascular Plants, but every year around 2,000 new species are added to the total. And we know even less about fungi: with only 148,000 species officially described, as many as 90% of fungal species remain unknown to science. Our scientists and partners dedicate much effort to accelerating the description of new species. Among the plants and fungi that they have scientifically described and named in recent years are a new species of wild tobacco, *Nicotiana insecticida*, that snares and kills small insects; a 'ghost' orchid, *Didymoplexis stella-silvae*, which grows in almost complete darkness; and a fungus from a group of species that cause the Panama disease that wipes out banana crops, *Fusarium odoratissimum*.

Kew's scientists are dedicated to unlocking the secrets of plants and fungi at a molecular and cellular level.

FINDING USEFUL PLANTS AND FUNGI

Despite there being 7,000 catalogued edible plant species on Earth (and likely many thousands more), just 15% of crop plants contribute to 90% of humanity's food intake, and more than four billion people rely on just rice, maize and wheat. Similarly, only six species produce 80% of global biofuel. Often, commercial crops are developed to thrive in particular environmental niches and may not be able to tolerate the higher temperatures and less predictable rainfall resulting from climate change. So, new crops are needed with sufficient genetic diversity to remain robust under variable weather regimes and against emerging pests and diseases. Kew is contributing to this work by studying and conserving the wild relatives of our current crops, along with overlooked and underutilised food and energy plants that could be developed for more widespread use.

One way in which our scientists gain a better understanding of which plants might be useful to us is through our Tree of Life initiative. This ongoing work aims to map the origins and evolution of all 13,600 individual plant genera and half of the 8,000 fungal genera, using our unrivalled collections. Once completed it will provide a 'biological roadmap' that can be used to chart evolutionary histories, understand the relationships between different species and predict where on the tree of life plants with useful traits might reside. It is already proving its value in this regard. For example, understanding the plant tree of life has helped to direct the search for new antimalarial compounds towards 'hot nodes' – clusters of species known to be used to prevent or treat malaria, and their close relatives.

As well as focusing on individual species, Kew applies its scientific expertise at broader scales to assess the health of ecosystems and identify ways in which biodiversity can help resolve challenges faced by society. Wakehurst – Kew's sister garden in the heart of the West Sussex countryside – is a particular focus for research into quantifying the wider values, services and benefits that biodiversity provides for people and the environment, as part of the 'Nature Unlocked' programme.

Over 2.4 billion seeds are stored at minus 20 degrees centigrade in the vault of the Millennium Seed Bank, where they can be accessed for research and restoration work.

USING OUR COLLECTIONS

The Millennium Seed Bank at Wakehurst underpins much of Kew's research in the fields of conservation, biodiversity and plant uses. Today, it houses seeds from more than 48,000 plant species collected in 190 countries and territories with our partner organisations. When wildfires razed 24 million hectares of vegetation in Australia, including a home of the rare clover glycine pea, *Glycine latrobeana*, the Millennium Seed Bank was able to send back previously collected seeds to generate plants for restoring the damaged habitat. The seed bank, along with the living collections at Kew Gardens and Wakehurst, also represents a priceless reserve of genetic material that can be mined for useful plant traits to reveal how a specific species has adapted to particular environments, and to signify how it may respond to future climate change. Gaining such understanding is enabling Kew scientists to find better ways to conserve plants and fungi, within and outside their natural ranges, and to identify species and molecules that can be used sustainably and equitably to benefit humankind.

Sharing Knowledge and Resources to Resolve Global Challenges

By using the latest technology to digitise and share our collections, educating a new generation of environmental experts, and building global networks of partners, Kew is laying the groundwork for Earth's natural resources to be managed sustainably in future.

The scale of today's major environmental challenges requires inputs from wide-ranging experts and organisations. For this reason, Kew is making its resources available more widely online. In late 2021, the UK Government committed £15 million of funding to support the digitisation and long-term preservation of our Herbarium and Fungarium collections. Every specimen in the collection is packed full of information – from where and when it was collected, to the weather and vegetation at the time, and how it was named. Comparing data from many specimens can shed light on changes to habitats and environmental conditions over time, such as the clearance of forests for agriculture. Once completed, the database will be freely available to researchers around the world, providing a valuable resource for understanding biodiversity loss and the effects of climate change.

We already share our resources, expertise and knowledge through our extensive teaching programme, and via collaborations and partnerships with experts and institutions globally. Seeking to equip the next generation with the skills and understanding to manage and use the world's resources sustainably, we welcome a total of 100,000 pupils to Kew Gardens and Wakehurst every year, and train numerous diploma, MSc and PhD students. At the same time, we work with a broad range of partners to achieve shared goals and objectives, from the sustainable use of plant-based products to delivering climate-positive initiatives. And by working with governments around the world, we ensure that the findings of our scientific research are used to underpin effective policies aimed at stopping biodiversity loss and restoring habitats. We aspire to help create a world where nature is protected, valued by all and managed sustainably.

Digitising herbarium specimens.

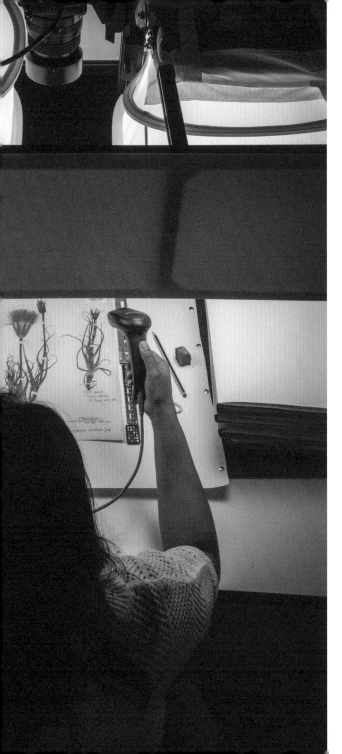

Schools, Community and Access Learning

Building understanding is the key first step to empower people to support and take action for conservation and sustainability.

We offer unique opportunities for millions of visitors to explore and understand the importance of plants and fungi, their relevance to our lives today and the future viability of the planet. Kew runs one of the world's largest botanic garden schools' programmes, with 100,000 school pupils, aged five to 18 years, visiting our sites annually and over 6,500 primary and secondary teachers accessing Endeavour, our online library of teaching resources.

Reaching out to our community is important; we work with partners across London to deliver a wide range of learning and engagement programmes for families and early years, youth and adults. We run a community access scheme offering annual membership to organisations that provide services to people who face physical, sensory, psychological or social barriers to visiting Kew.

More broadly, Kew's national outreach programme, Grow Wild, inspires tens of thousands of people across the UK to grow and value native wild plants. Through grants, seed and fungus kits and digital engagement, Grow Wild has reached communities in 30% of the most deprived areas of the UK.

Visit Wakehurst

Explore Kew's wild botanic garden at Wakehurst, home to the Millennium Seed Bank, diverse landscapes and plants from across the globe in West Sussex. Set in the dramatic landscape of the High Weald, Wakehurst's 188 hectares (465 acres) include an American prairie, ornamental and water gardens, temperate woodlands, the Loder Valley Nature Reserve and an Elizabethan mansion.

Many important plant collections thrive in the higher rainfall and moister soils – especially from eastern Asia, South America, Australia and New Zealand. Find out more at www.kew.org/wakehurst

Further Reading

Antonelli, Alexandre. (2022). *The Hidden Universe: Adventures in biodiversity.* Ebury in association with Kew.

Brereton, Catherine and McGuiness, Jane. (2018). *Kew Children's Guide.* Bloomsbury in association with Kew.

Bynum, Helen and William. (2014). *Remarkable Plants That Shape Our World.* Thames & Hudson in association with Kew.

Desmond, Ray. (2007). *The History of the Royal Botanic Gardens, Kew,* 2nd edition. Royal Botanic Gardens, Kew.

Edwards, Ambra. (2021). *The Plant Hunter's Atlas.* Greenfinch in association with Kew.

Farjon, Aljos. (2022). *Ancient Oaks in the English Landscape,* 2nd edition. Royal Botanic Gardens, Kew.

Flanagan, Mark and Kirkham, Tony. (2009). *Wilson's China: A century on.* Royal Botanic Gardens, Kew.

Fry, Carolyn; Seddon, Sue and Vines, Gail. (2011). *The Last Great Plant Hunt: The story of Kew's Millennium Seed Bank.* Royal Botanic Gardens, Kew.

Griggs, Patricia. (2011). *Joseph Hooker: Botanical trailblazer.* Royal Botanic Gardens, Kew.

Harrison, Christina. (2019). *Kew's Big Trees,* 2nd edition. Royal Botanic Gardens, Kew.

Harrison, Christina and Gardiner, Lauren. (2016). *Bizarre Botany.* Royal Botanic Gardens, Kew.

Harrison, Christina and Kirkham, Tony. (2019). *Remarkable Trees.* Thames & Hudson in association with Kew.

Harrison, Christina; Rix, Martyn and Yamanaka, Masumi. (2015). *Treasured Trees.* Royal Botanic Gardens, Kew.

The Hive at Kew. (2016). Royal Botanic Gardens, Kew.

Ikin, Ed. (2021). *Rare Plants.* Welbeck in association with Kew.

Kew Pocketbooks series (2020–2023). Royal Botanic Gardens, Kew.

Linford, Jenny. (2022). *The Kew Gardens Cookbook.* Royal Botanic Gardens, Kew.

North, Marianne. (2018). *Marianne North: The Kew Collection.* Royal Botanic Gardens, Kew.

Parker, Lynn and Ross-Jones, Kiri. (2013). *The Story of Kew Gardens in Photographs.* Arcturus Publishing.

Payne, Michelle. (2018). *The Temperate House.* Royal Botanic Gardens, Kew.

Payne, Michelle. (2016). *Marianne North: A very intrepid painter,* 2nd edition. Royal Botanic Gardens, Kew.

Price, Katherine. (2017). *Get Plants: how to bring green into your life.* Royal Botanic Gardens, Kew.

Rix, Martyn. (2021). *Indian Botanical Art.* Roli Books in association with Kew.

Scott, Katie and Gaya, Ester et al. (2019). *Fungarium.* Big Picture Press in association with Kew.

Scott, Katie and Kirkham, Tony (2022). *Arboretum.* Big Picture Press in association with Kew.

Scott, Katie and Willis, Kathy. (2016). *Botanicum.* Big Picture Press in association with Kew.

Sherwood, Shirley. (2019). *The Shirley Sherwood Collection: Masterpieces of botanical art.* Royal Botanic Gardens, Kew.

Sherwood, Shirley and Rix, Martyn. (2008). *Treasures of Botanical Art.* Royal Botanic Gardens, Kew.

Teltscher, Kate. (2020). *Palace of Palms.* Picador in association with Kew.

The Science of Plants. (2022). Dorling Kindersley in association with Kew.

Walker, Kim and Nesbitt, Mark. (2019). *Just the Tonic: A natural history of tonic water.* Royal Botanic Gardens, Kew.

Wilford, Richard and Willoughby, Sharon. (2020). *The Agius Evolution Garden.* Royal Botanic Gardens, Kew.

Wilford, Richard. (2017). *The Great Broad Walk Borders.* Royal Botanic Gardens, Kew.

Willis, Kathy and Fry, Carolyn. (2014). *Plants from Roots to Riches.* John Murray, in association with Kew.

Acknowledgements

All photography by Jeff Eden and Ines Stuart-Davidson except where specified below.

We would like to thank the following for providing photographs and for permission to reproduce copyright material:

Cover: front, Ines Stuart-Davidson, back, Jeff Eden.
Inside: Jeff Eden, Ines Stuart-Davidson, John Millar, Tomás del Amo, Jean Postle, Alberto Trinco, Tiziana Ulian, Wolfgang Stuppy, Wilkinson Eyre Architects, Marks Barfield Architects, Walters & Cohen Architects, John Pawson Ltd, Skyvantage, Historic Royal Palaces.

Thank you to the following for their valued help and advice with this edition: Fiona Ainsworth, Richard Barley, Sandra Botterell, Elinor Breman, Maxine Briggs, Colin Clubbe, Paul Denton, Maria Devaney, Richard Deverell, Hélèna Dove, Janine Durkin, Aisyah Faruk, Carolyn Fry, Gina Fullerlove, Tony Hall, Brie Langley, Kevin Martin, Mark Nesbitt, Rhian Smith, Adam Thow, Alberto Trinco, Richard Wilford, Julia Willison, Colin Ziegler, and the Kew Press Office team.

Index

Reprinted with revisions in 2017, 2018, 2019, 2020, 2023
First published in 2014 by Royal Botanic Gardens, Kew, Richmond, Surrey, TW9 3AE, UK
www.kew.org

ISBN 978 1 84246 786 2

Distributed on behalf of the Royal Botanic Gardens, Kew in North America by the University of Chicago Press, 1427 East 60th Street, Chicago, IL 60637, USA.

British Library Cataloguing in Publication Data
A catalogue record for this book is available from the British Library.

Text: Katherine Price
Design, typesetting and page layout: Jeff Eden
Project and production management: Georgina Hills
Revisions: Maxine Briggs, Paul Denton, Maria Devaney, Carolyn Fry, Gina Fullerlove, Christina Harrison, Mark Nesbitt, Rhian Smith, Julia Willison
Front and back cover photographs: Ines Stuart-Davidson, Jeff Eden

For information or to purchase all Kew titles, please visit shop.kew.org or email publishing@kew.org

Kew's mission is to understand and protect plants and fungi, for the wellbeing of people and the future of all life on Earth.

Kew receives approximately one third of its funding from Government through the Department for Environment, Food and Rural Affairs (Defra). All other funding needed to support Kew's vital work comes from members, foundations, donors and commercial activities, including book sales.

Printed in Italy by Graphicom